Plumbing

PLUMBING

Roy Treloar

Colchester Institute

b

Blackwell
Science

© 1994 by
Blackwell Science Ltd
Editorial Offices:
Osney Mead, Oxford OX2 0EL
25 John Street, London WC1N 2BL
23 Ainslie Place, Edinburgh EH3 6AJ
238 Main Street, Cambridge,
    Massachusetts 02142, USA
54 University Street Carlton,
    Victoria 3053, Australia

Other Editorial Offices:
Arnette Blackwell SA
1, rue de Lille
75007 Paris
France

Blackwell Wissenchafts-Verlag GmbH
Kurfürstendamm 57
10707 Berlin
Germany

Blackwell MZV
Feldgasse 13
A-1238 Wien
Austria

First published 1994
Reprinted 1995

Set by Florencetype Ltd, Kewstoke, Avon
Printed and bound in Great Britain by
Bell & Bain Ltd, Glasgow

DISTRIBUTORS

Marston Book Services Ltd
PO Box 87
Oxford OX2 0DT
(*Orders*: Tel 01865 791155
         Fax: 01865 791927
         Telex: 837515)

USA
Blackwell Science, Inc.
238 Main Street
Cambridge, MA 02142
(*Orders*: Tel: 800 215-1000
         617 876-7000
         Fax: 617 492-5263)

Canada
Oxford University Press
70 Wynford Drive
Don MIlls
Ontario M3C 1J9
(*Orders*: Tel: 416 441-2941)

Australia
Blackwell Science Pty Ltd
54 University Street
Carlton, Victoria 3053
(*Orders*: Tel: 03 347-5552)

A catalogue record for this book is available from
the British Library.

ISBN 0-632-03761-X

Library of Congress
Cataloging in Publication Data

Treloar, Roy.
    Plumbing/Roy Treloar.
    p. cm.
    Includes index.
    ISBN 0-632-03761-X
    1. Plumbing. 2. Plumbing—Examinations,
    questions, etc.
    3. Plumbing—Problems, exercices, etc. I. Title
TH6123.T74 1994
696'. 1—dc20   94-13670
                                           CIP

Extracts from British Standards are reproduced with the permission of BSI. Complete copies can be obtained by post from BSI Sales, Linford Wood, Milton Keynes, MK14 6LE.

Contents

Contents

Introduction

This book is designed to provide easily accessible information on a wide range of plumbing subjects. Few books could possibly hope to have all the answers to questions relating to plumbing skills; I have, however, endeavoured to cover as many topics as possible in the hope that the book will be a source of useful information both for the student with no knowledge whatsoever of the subject, and the trained plumber seeking guidance in particular areas of study.

Covered here are topics found in NVQs, levels 2 and 3, these levels providing the current NVQ for plumbing in the UK; additional skills are identified and these are of a sort which no respectable plumber could possibly ignore.

The book is in ten parts. Parts 1 to 9 offer a programme of training and information while Part 10 is designed to allow you to assess your level of knowledge. There are sections for self- and supplementary assessments and a few typical plumbing problems.

Broadly speaking, the subject matter dealt with in the supplementary assessment questions is introduced in the same order as the subject matter of the book itself; hence these sections can be used as self-learning packages.

Further useful information can be found in the preliminary pages which follow. The nature of the NVQ in plumbing is, for example, identified, as is a guide showing how to complete the scheme.

Special note to college lecturers

Answer books (providing answers to the supplementary questions and problems identified on pages 341–390) are available to colleges and training centres. These may be purchased by sending a cheque for £3.50 (inc. postage and packing) to the following address: School of Construction, Colchester Institute, Sheepen Road, Colchester, Essex, CO3 3LL. Cheques should be made payable to R. Treloar.

List of Abbreviations

BSP   British standard pipe thread
c.h.   central heating
cpc   circuit protective conductor
dhw   domestic hot water
dpm   damp proof membrane
f & e   feed and expansion
f & r   flow and return
LPG   liquefied petroleum gas
PTFE   polytetrafluoroethylene
PVC   polyvinylchloride
wg   water gauge

National Vocational Qualifications

National vocational qualifications (NVQs) are worked-based awards which have replaced the City & Guilds craft and advanced craft certificates in plumbing. They have been introduced in order to:

☐ Allow the candidate the opportunity to gain the qualification at his/her own pace.
☐ Permit the candidate to carry forward from one job qualification to another any appropriate competence that has common applications.
☐ Permit mature candidates with considerable work experience in plumbing to submit details of work completed, thus enabling them to receive accreditation of prior learning.
☐ Allow individuals the opportunity to be accredited for work completed on site.
☐ Indicate to employers that candidates are competent in the level to which they have been awarded, i.e. they have actually carried out the tasks indicated and not simply completed an assignment or test on the subject.

There are various levels of NVQs, levels 2 and 3 being most applicable to plumbing trainees.

☐ *Level 2* indicates competence in a significant range of varied work activities, performed in a variety of ways. Some of the activities may involve many parts or be non-routine. Individual responsibility and decision-making is required at times.
☐ *Level 3* indicates competence in a broad range of varied work activities covering a wide variety of contexts, most of which are complex or complicated. There is considerable responsibility and control or guidance of others is often required.

NVQs differ from the previous City & Guilds assessment system, where end examinations were the ultimate goal, in that assessment is ongoing throughout the training period. Candidates have to provide evidence of competence which may include such things as witness testimonials or written declarations from employers (e.g. customers, foremen, the plumber in charge, etc.); photographs; informal and formal tests, both written and practical; these are considered by the NVQ assessor, who must be satisfied, after weighing up all the evidence, that a candidate is competent to carry out the work he claims to be able to do.

In order to achieve the award at level 2 the trainee plumber has to complete eight units which in turn are broken down into individual elements. A unit of competence is the smallest unit worthy of separate accreditation. The award at level 3 equally has eight units and is subdivided into elements.

Commencement of an NVQ Qualification in Plumbing

Many colleges of further education carry out two roles to:

(1) Act as a training provider offering a plumbing course in its entirety which certainly will include education and training to the standards required for NVQ assessment.
(2) Act as an approved NVQ assessment centre where an assessment is made as to your level of competence.

Upon enrolment on a course of training the college will advise you as to the correct procedure for completion of the scheme. Generally after an initial induction period your centre will register you with the Awarding Body who will in turn provide you with the following items and documentation:

☐ A Candidate Activity Log Book,
☐ Several Activity Record Sheets,
☐ A Record of Assessment (level 2 or 3),
☐ A Storage File Box.

Initially candidates would register for level 2 and follow this by taking level 3; however, they may register directly for level 3 as this covers all the competences covered in level 2 plus additional more complex duties. (Currently the sheetwork units are not in level 3.) Those who first achieve level 2 simply carry across those assessments achieved into level 3. Guidance should be sought from the training centre to ensure correct enrolment onto the required training programme.

Each of the items provided by the awarding body are designed to help you complete a portfolio of evidence showing how you have achieved all the specified criteria as laid down in the assessment documentation. The evidence is kept within the box file as necessary.

Activity Logbook and Record Sheets

These are designed to assist you in keeping a self-assessment record of the actual work situations you have experienced. Generally speaking, upon completion of a task you would complete an *Activity Record Sheet*, as shown on page xiv. As much information as possible must be entered on this sheet, which allows you to tick off any boxes showing tasks you feel competent to carry out.

The Activity Record Sheet will need to be signed by the work supervisor or employer to indicate its authenticity; there is also provision in the logbook to testify that this, too, is an accurate account of the work done. These entries, along with any photographs or statements from customers, etc., are the evidence of competence.

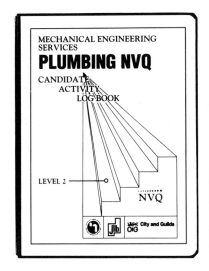

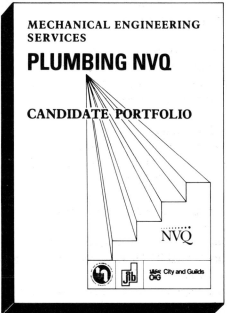

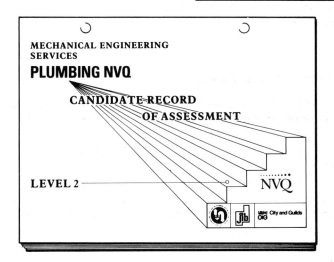

MES PLUMBING NVQ   **ACTIVITY RECORD SHEET**

NAME OF CANDIDATE

B. GOODRIDGE

NVQ LEVEL   2

UNIT TITLE   VARIOUS

JOB NUMBER   5.

> TO COMPLETE THE INSTALATION
> OF A SECTION OF CAST IRON
> DRAINAGE.

This record should include details of any work that is appropriate to the unit. It should list the activities, detail any problems stating how these were overcome, describe any checks carried out for compliance with regulations or company policy, describe planning activities e.g. liaison with other trades etc., provide details of any safety precautions and list materials and equipment used.

N.B. Sketches where appropriate should be provided to supplement written notes and be included in your portfolio of evidence.

PLEASE INDICATE THE PAGE NUMBERS OF YOUR LOG BOOK TO WHICH THE WORK DETAILED ON THIS SHEET RELATES.

DATE(S) AND LOCATION(S) OF WORK ACTIVITIES

15TH JAN 1994   COLCHESTER INSTITUTE; (SITE SIMULATION).

	PAGE Nº	ELEMENT
1. PREPARED A LIST OF MATERIALS & WITHDREW FROM STORE.	8	1·3
2. CHECKED MATERIALS CONFORMED TO SPECIFICATION.	8	1·3
3. COMPONENTS CHECKED FOR DAMAGE & CONFORMING TO CLIENTS REQUIREMENTS.	8	1·3
4. SELECTED CORRECT TOOLS & ENSURED THEY WERE IN A SAFE CONDITION.	8	1·3
5. SET OUT PIPE RUN TO REQUIRED GRADIENT & SUPPORTED AS NECESSARY USING 'PEA' SHINGLE.	8	1·3
6. JOINTED TO CAST IRON 100 DIA PIPE USING NON-MANIPULATIVE TYPE COMPRESSION JOINTS (TIMESAVER).	7	1·3
7. COMPLETED AIR TEST TO SYSTEM IN ACCORDANCE TO B.S. 8301.	10	1·4
8. CARRIED OUT WATER TEST AS AN ADDITIONAL TEST TO CONFIRM 1·5m WATER TEST WOULD HOLD.	10	1·4
9. CHECKED SYSTEM COMPONENTS INSTALLED AS PER SPECIFICATION.	13	2·1
10. FAULTS IDENTIFIED & RECTIFIED AS NECESSARY	13	2·1
11. CARRIED OUT TEST FOR ALIGNMENT & OBSTRUCTION USING MIRROR & TORCH AND 'BALL'.	16	2·2
12. RECORDED TEST RESULTS ONTO SCHEDULE.	13	2·1
PLANNING:		
* LIAISON WITH BRICKLAYER TO ARRANGE SUITABLE TIME TO WORK IN TRENCH, WHO ALSO HAD WORK TO COMPLETE WITHIN THE SAME AREA.	40	5·3
LEGISLATION:		
* CHECKED WORK COMPLIED WITH BUILDING REGULATIONS, PART 'H', TO ENSURE CORRECT BEDDING.	13	2·1
* COMPLETED WORK IN ACCORDANCE TO B.S. 8301.	13	2·1
SAFETY:		
* OBSERVED ALL SAFETY ASPECTS WHILST WORKING IN TRENCH — IE: WORE PROTECTIVE CLOTHING EG HARD HAT & BOOTS.	27	4·1
* MADE AREA SAFE FROM PEOPLE FALLING INTO HOLE	34	4·4
PROBLEM SOLVING:		
NO PROBLEMS ENCOUNTERED.		

Candidate signature: B Goodridge.

Date: 19TH JAN 1994

Continuation sheets should also be signed and dated.

PERSON OVERSEEING ACTIVITY: R. TRELOAR   SIGNATURE: DATE: 20TH Jan 94.

Typical Completed Activity Record Sheet

Commencement of an NVQ Qualification in Plumbing

UNIT 1 INSTALL AND TEST THE COMPONENTS OF THE SYSTEM
ELEMENT 1.3 Fabricate, position and fix components

TYPES OF EVIDENCE

PERFORMANCE

Assessment method i: Self Assessment on at least one occasion for the bending of materials, as scheduled below

	Copper			LCS
	8mm	15mm	28mm	1" BSP

Bending

- hand
- machine

Assessment method ii: Self assessment on at least one occasion for the jointing of materials, as scheduled below (abbreviations used: CI = cast iron; C = copper; LCS = low carbon steel; PP = pressure pipe; SW = soil and waste systems)

	C (mm) 100	C (mm) 8 15 28	LCS (inches BSP) 1 2	Plastic PP (mm) 15	SW (mm) 100

Jointing

soldered
- soft
- hard

	CI (mm) 100	C (mm) 8 15 28	LCS inches BSP 2	Plastic PP (mm) 15	SW (mm) 100

compression
- manipulative
- non-manipulative
- push fit
- solvent

threaded
- hand
- machine
- flanged

7

This shows the work completed as per the activity record sheet opposite.
A tick may simply be used or, as shown here, by the job number thus making for easy cross referencing.

...ment on at least one occasion for
...allations and components as follows
...o c) as appropriate).

a   b   c

AND EACH OF THE FOLLOWING

3 above ground discharge pipework and sanitation systems
4 below ground drainage systems
5 gas supply
6 electrical systems

Assessment method iv: Self assessment on at least one occasion for small commercial, industrial or public buildings as follows, for pipes of plastic, LCS and copper up to 54mm

EITHER group 1, above
OR group 2, above
AND group 5, above

a) Components are checked for damage confirm to specification and schedule
b) Components, materials and equipment chosen are correctly set out, assembled, fabricated and installed
c) Tools and equipment are correctly selected and safely used

SUPPLEMENTARY

Knowledge covered: Choice of materials, fixing devices, connections to and deficiencies of services and action to take, why components should be fixed and conform to codes of practice and regulations, alternative fixing methods for different building fabrics, work methods, cutting, potential dangers arising from site conditions and action to be taken

8

Extract from the Candidate Activity Log Book

You will find that the installation work you undertake does not occur in a neat sequence and as a result you will probably have a number of activity record sheets 'on the go' at the same time. Any work situations which you are unlikely to complete on site will be covered by the training establishment; these college-based activities will equally require the production of evidence.

When you have completed all the self-assessment for a particular element you should present your evidence to your NVQ assessor who will then decide what, if any, further evidence you will need to provide to prove your competence.

Supplementary assessments and additional evidence

Where evidence is submitted to the assessor the additional assessment required may be in the form of oral, written or practical testing in each case so that those responsible can satisfy themselves as to the competence of the individual.

Candidate record of assessment

The *Record of Assessment* is the 'master' document which provides full details of the standards, performance and supporting knowledge requirements. As you submit evidence to the NVQ assessor of your competence to meet the requirements of the different units, providing the assessor is satisfied as to your level of knowledge, the ticked boxes will be transferred by the assessor to this record book and countersigned as necessary. It is when this Record of Assessment is completed that you will be issued with the final NVQ certificate appropriate to your level.

 **City and Guilds** of London Institute

National Vocational Qualification

MECHANICAL ENGINEERING SERVICES – PLUMBING
LEVEL 2

**This Certificate
is awarded to**   MICHAEL OWEN

SPECIMEN

**The holder has one or more formal Certificates of Unit Credit
by which this Certificate was earned**

Awarded OCTOBER 1993   9310/012345/600901/ABC0009/1/00/00/00

*Director-General
City and Guilds of London Institute*

*Chairman
The Joint Industry Board for
Plumbing Mechanical Engineering
Services in England and Wales*

*Chairman
Education and Training Committee
National Association of Plumbing
Heating and Mechanical
Services Contractors*

The City and Guilds of London Institute is incorporated by Royal Charter and was founded in 1878

N537

Part 1

Underpinning Knowledge

The Plumbing Industry

The role of the plumber

In the eyes of many lay people the plumber is simply someone who joins pipes together, running the water from one appliance or another. Few people appear to give much consideration to the depth of knowledge required to carry out the basic plumbing skills, or the trained professional's need to have a full understanding of related activities – or his ability to design installations and identify likely problems before they occur.

Unfortunately many students come to college and find it difficult to take on board skills associated with tasks which their own company does not undertake. They seem to think that skills they are not making use of now are irrelevant. I would respond by saying that if you want to have greater freedom and be in high demand – and not only by your present employer – then, yes, these extra skills are very relevant indeed.

Trained plumbers are able to turn their hands to many, if not all, the following skills, although levels of knowledge will vary from one individual to another:

- [ ] The supply and distribution of cold and hot water for drinking purposes, sanitation, heating and fire fighting, etc., and the connection of associated equipment and controls.
- [ ] The removal of water from the building via a suitable system of drainage, to include foul water from soil and waste appliances and surface water from roofs and paved areas.
- [ ] The weathering of roof penetrations, etc., in metallic sheet materials.
- [ ] The supply and provision of fuels, including gas, oil, solid fuel and electricity, to various appliances and components, such as those for heating or cooking, and the provision of such ventilation as is necessary for combustion.
- [ ] The removal of the combustion products from appliances by way of a safe and effective flue system.
- [ ] Designing and estimating the cost of any of the above installations, in a domestic situation, giving efficient and effective usage.
- [ ] The knowledge to identify, rectify and service any of the said installations.

In addition plumbers are well served by a number of skills normally associated with other operatives, such as carpenters and bricklayers – skills that might enable them, for example, to make good a hole or remove/replace a floorboard, as part of the job.

The impressions you create are very important; going to work in dirty clothes and trainers gives the same impression as submitting an estimate for a job on a piece of paper taken from a school notebook. Laying a dustsheet down, even for the smallest job, takes very little time, yet it gives a lasting effect. A positive image projected today is likely to bring you work for tomorrow.

Plumbing Organisations
(addresses correct at time of going to press)

There are many organisations which have connections with the plumbing industry, the following identifying a small collection representing the Mechanical Engineering Services, Plumbing sector:

National Association of Plumbing Heating & Mechanical Services Contractors (NAPH & MSC)

Address:	Ensign House Ensign Business Centre Westwood Way, Coventry CV4 8JA
Tel:	0203 470626

Role: The employer organisation for the MES Plumbing Industry in England and Wales.

Scottish & Northern Ireland Plumbing Employers Federation (SNIPEF)

Address:	2 Warker Street Edinburgh EH3 7LB
Tel:	031 225 2255

Role: The employer organisation for the MES Plumbing Industry in Scotland and Northern Ireland.

British Plumbing Employers Council (BPEC)

Address:	Same as NAPH & MSC and SNIPEF

Role: The industry training organisation and lead body for plumbing.

The Joint Industry Board for Plumbing MES in England & Wales (JIB for PMES)

Address:	Brook House Brook Street, St Neots PE19 2HW
Tel:	0480 476925

Role: Deals with the grading of plumbing operatives and registration of apprentices. Also agrees terms and conditions of employment.

Institute of Plumbing (IOP)

Address:	64 Station Lane Hornchurch, Essex RM12 6NB
Tel:	0708 472791

Role: An independent body having the prime objective of promoting better plumbing practices. The IOP operates a plumbing registration scheme.

Council of Registered Gas Installers (CORGI)

Address:	4 Elmwood Chineham Business Park Crockford Lane Basingstoke RG24 0WG
Tel:	0256 707060

Role: To register and verify the competence of those working within the gas industry.

Electrical, Electronic, Telecommunications & Plumbing Union (EEPTU)

Address:	Hayes Court West Common Road Hayes Bromley BR2 1PT
Tel:	071 215 0609

Role: The trade-union representing those in the MES-Plumbing industry.

Safety at Work

Health and Safety at Work Act 1974

This is the Act of Parliament which provides the framework for all future safety legislation. The Act involves everyone: employers; employees; the self-employed; managers; representatives; manufacturers, etc. in matters of health and safety. Failure to meet the requirements of this Act, together with other health and safety regulations carried out under the provisions of this Act, constitutes a criminal offence and prosecution by the Health and Safety Executive (HSE).

One of the main features of this Act is to make everyone responsible for site safety, thus ensuring the safety of themselves and others around. Employers must provide and maintain plant, make arrangements for safety, information, instruction, training and supervision, etc., as necessary, to ensure the health and safety of their employees.

There are some 50 to 60 Acts and over 400 sets of Regulations relevant to Health and Safety at Work. It is these Regulations which identify the minimum standard for safety required by law.

Safety signs

Many signs are used on a construction site and their meaning needs to be considered and understood. Signs are not meant for show but are placed there for a specific reason. The classification of a sign will fall into one of the following categories:

- ☐ *Information* square or rectangular, white on green background
- ☐ *Warning* triangular, black on yellow background
- ☐ *Mandatory* circular, white on blue background
- ☐ *Prohibition* circular, red border and cross bar on white background

Fire prevention

The 1961 Factories Act, among other Acts, provides for the appropriate means of fire-fighting within premises. Generally the most important point to consider is that you have made suitable provisions to deal with an outbreak of fire should one become established, in fact, it is likely your insurance cover would be invalid if you had not done so. Always check that nothing is left smouldering on completion of the work. Gas and electrical supplies should be turned off as quickly as possible and combustible materials moved out of the reach of the fire.

Should a fire break out it is essential to use the correct equipment to control the spread of fire as using the wrong type of extinguishing agent can make things worse.

Fire extinguisher selection chart

Type of fire	Water	Foam	CO$_2$	BCF*	Powder	Fire blanket
Wood, paper and textiles, etc.	yes	yes	no	yes	no	yes
Flammable liquids – oil and paints, etc.	no	yes	yes	yes	yes	yes
Electrical equipment	no	no	yes	yes	yes	no
Extinguisher colour	red	cream	black	green	blue	red

* Bromochlorodifluoromethane

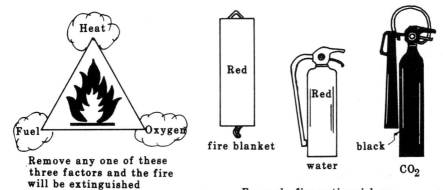

Remove any one of these three factors and the fire will be extinguished

fire blanket   water   CO$_2$

Example fire extinguishers

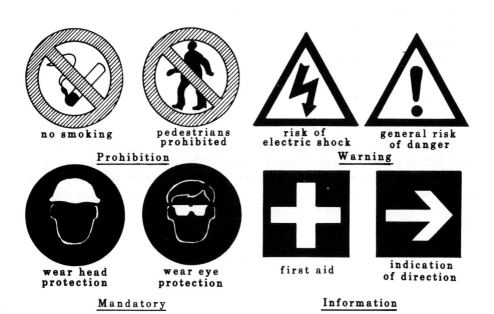

no smoking   pedestrians prohibited   risk of electric shock   general risk of danger

Prohibition   Warning

wear head protection   wear eye protection   first aid   indication of direction

Mandatory   Information

Safety at Work

Protective Clothing

Sooner or later you will need to use protective clothing and equipment. Overalls and safety footwear may be provided by the employer, or you may have to buy these items yourself. Employers have a legal duty to provide all other protective equipment free of charge and the employee must use it correctly and report any defect or damage. Visitors to the site or other workers are also entitled to the same protection.

The Construction (Head Protection) Regulations 1989

The law requires the use of suitable head protection on all building sites unless there is no risk of head injury other than by the person falling. Safety hats should be adjusted to fit correctly; your failure to make the correct adjustment may mean that you are not providing the necessary level of safety.

The Construction (Protection of Eyes) Regulations 1974

Safety glasses, goggles or eye shields must be worn when there is any foreseeable risk of eye injury. Eye injury can result from:

☐ The use of power tools   (drilling, grinding and threading)
☐ Hammering and driving tools   (cutting, chipping and chiselling)
☐ Flying particles   (dust and chemical splashes)
☐ Welding processes   (sparks and molten splashes)
☐ Glare from light   (electric ark welding)

The Noise at Work Regulations 1989

Sound levels are measured in decibels (dB). Employers are required, as far as possible, to keep noise levels below 85 dB; where over 90 dB is experienced suitable ear protectors must be provided and the work area designated an ear protection zone.

Protection of skin and hands

Your skin, and particularly your hands, should be protected not only from cuts and abrasions but also from materials and substances which on contact can lead to infection. Skin conditions such as industrial dermatitis and skin cancer can be the result of neglect.

The two main types of gloves available are: (1) hide, leather or similar materials, for damage due to roughness, heat, etc.; and (2) PVC, rubber or Neoprene to prevent damage due to contact with chemicals, cement, oils, etc.

Barrier creams can be used to give a minimum protection. At all times wash your hands to prevent illness due to ingesting toxic substances such as lead and copper, which may be impregnated onto your hands.

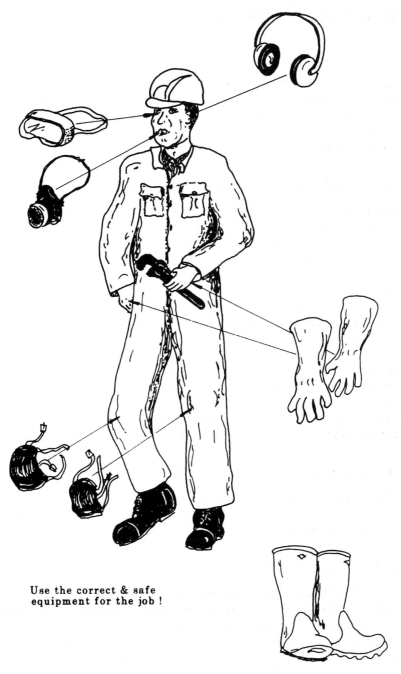

Use the correct & safe
equipment for the job !

Protective Clothing

Safety on Site

The vast majority of safe working practices are common sense. For example, do not lift extremely heavy loads; and when lifting, approach the load squarely and lift keeping a straight back, using the legs *not* the back and keeping the load close to the body. When lowering the load simply reverse the process. Many people today suffer with bad backs due to foolish acts when they were younger.

Hand tools should always be maintained in a safe condition and only used for their designed purpose.

Ladder work When using ladders check them over to see that they are in good condition and secure them to the stiles (not the rungs) while in use. Stand them on a firm even base and always use them at the correct angle of approximately 75° (ratio: 4 up to 1 out); above all never over-reach when working at high levels.

Working platforms Take particular care when working from a platform such as that provided by scaffolding; always consider those below. The scaffold should be erected only by certified operatives, and it should be checked weekly or after adverse weather conditions: thus its condition should be sound; but always have a visual inspection to identify obvious defects and above all never alter or work from ineffective scaffolding platforms. Defects to look for include:

☐ *Missing components*, e.g. toe boards or guard rails,
☐ *Poor assembly*, e.g. loose, overlapping or protruding boards,
☐ *Damaged scaffolding*, e.g. split boards and bent or rusty poles,
☐ *Unstable scaffolding*, e.g. no bracing, no tie-ins and no base plates,
☐ *Obstructed or overloaded scaffolding*.

Fragile roof coverings When working from roofs of this nature take extreme care; always use crawling boards, etc., to spread the load and never work directly off the roof covering itself.

Excavations Two types of accidents can occur with excavations: either the trench itself collapses or people fall into the trench. Therefore always make sure the trench is well supported, keeping heavy loads away from its edges; and erect barriers around the excavation where necessary. When using liquefied petroleum gas (LPG), such as propane, never leave the bottles in or around the trench as if the gas were to leak it would fall (being heavier than air) and fill the trench and an explosion may result.

Accidents should only occur in unforeseen circumstances: unfortunately most can be 'foreseen' only after the event. Most people think untoward accidents will not happen to them or that they are not at fault. Anyone who sees a hazard and does nothing about it is making it much more likely that an accident will occur. Make safe or report dangerous situations.

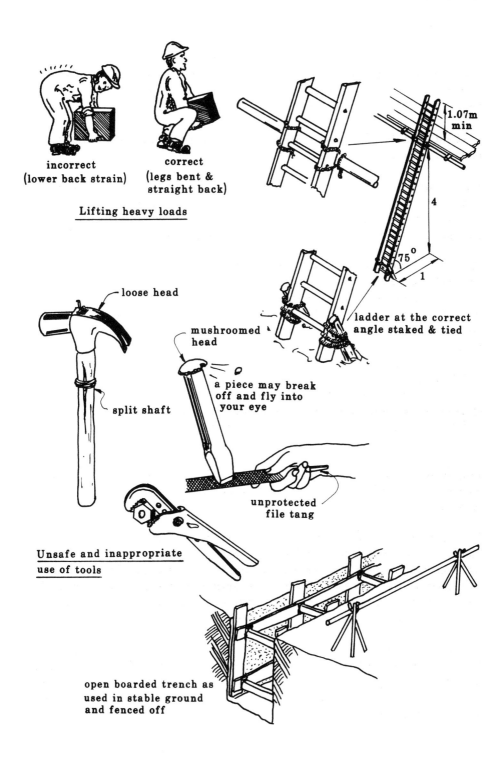

incorrect
(lower back strain)

correct
(legs bent &
straight back)

Lifting heavy loads

1.07m min

4

75°

1

ladder at the correct
angle staked & tied

loose head

split shaft

mushroomed
head

a piece may break
off and fly into
your eye

unprotected
file tang

Unsafe and inappropriate
use of tools

open boarded trench as
used in stable ground
and fenced off

Safety on Site

Regulations Governing Plumbing Work

The principal laws concerning plumbing works are the result of various Acts of Parliament, including the Water Act 1945; Gas Act 1948; Electricity Act 1936 and The London Building Act 1930. These have resulted in the following 'Statutory' Regulations being implemented:

Local Water Byelaws 1989 The local water byelaws set down the requirements to be observed to prevent wastage, undue consumption, misuse or contamination of water. The Local Water Authority administers these byelaws, which are the statutory requirements for the district and must be observed at all times on matters of water supply.

Gas Safety (Installation and Use) Regulations 1984 These regulations are concerned with work on gas fittings, including pipework, appliances, ventilation and the extract of flue gases, etc., so as to ensure the public is protected from any dangerous situations which may arise. The latest 1990 amendment (clause 3.3) prohibits any people employed from carrying out gas work unless they are members of a class approved by the Health and Safety Executive (i.e. CORGI registered).

The Electrical Supply Regulations 1988 and The Electricity at Work Regulations 1989 These regulations apply to all electrical equipment and installations. They require that in the event of an electrical fault the supply must not give rise to danger and be isolated, or cut off. It is now illegal to work on live electrical systems unless there is no other way in which the work can be done. Compliance with the Institution of Electrical Engineers (IEE) Wiring Regulations will, in general, satisfy the requirements of these Statutory Regulations in force although the IEE Regulations themselves are not statutory.

The Building Regulations 1991 Building requirements differ, depending where in Britain the work is carried out, but in general the Building Regulations apply. These are administered by the local authority via the building control officer. The Building Regulations contain no technical detail, these being found in a series of approved documents designated A to M (e.g. H1 deals with sanitary pipework and drainage). The purpose of the Building Regulations among other things is to conserve fuel and secure the welfare, health, safety and convenience of people in or around buildings.

British Standards These are not statutory documents although they are constantly referred to by those seeking to meet the requirements of the law. They are methods of design and installation practices which should be maintained whenever possible.

Regulations Governing Plumbing Work

Top left Building Regulations.
Top right Electricity Regulations.
Middle Water Byelaws.
Bottom left British Standards.
Bottom right Gas Safety Regulations.

Identification of Pipework

Relevant British Standards
BS 1192 and BS 1710

In order to understand pipework drawings one needs to be able to recognize a system of drawing symbols which is standard across the country; this prevents unnecessary labelling and assists in the clarification of various details. The British Standards Institution has produced a series of symbols which is consistently used; some different symbols do, however exist, these usually being ones used prior to current British Standards or because no BS symbol can be found.

Sometimes one is faced with a series of pipes running along the wall of, for example, a boiler room, and the nature/contents of each would soon be forgotten if the pipework were not labelled. One could simply write the name of the pipe contents on the pipe or its lagging, but this might well be somewhat time consuming and difficult to do neatly. Therefore a system of colour coding has been designed, again conforming to a British Standard, to enable pipe contents identification.

Pipe contents identification chart

Pipe contents	Basic colour*		Specific colour†		Basic colour*
Untreated water	Green		Green		Green
Drinking water	Green		Auxiliary blue		Green
Cold down service	Green	White	Blue	White	Green
Hot water supply	Green	White	Crimson	White	Green
Central heating <100°C	Green	Blue	Crimson	Blue	Green
Boiler feed	Green	Crimson	White	Crimson	Green
Chilled	Green	White	Em. green	White	Green
Fire extinguishing	Green		Red		Green
Condensate	Green	Crimson	Em. green	Crimson	Green
Steam	Silver grey		Silver grey		Silver grey
Natural gas	Yellow ochre		Yellow		Yellow ochre
Diesel fuel oil	Brown		White		Brown
Furnace fuel oil	Brown		Brown		Brown
Compressed air	Light blue		Light blue		Light blue
Drainage	Black		Black		Black
Electrical conduits	Orange		Orange		Orange

* 150 mm (see bottom of figure opposite); † 100 mm (see bottom of figure opposite)

Occasionally it will be necessary also to indicate the direction of flow; this is shown by an arrow situated in the proximity of the pipe contents identification code. With central heating pipework the word *flow* or the letter *F* is shown on one pipe and *return* or *R* on the other.

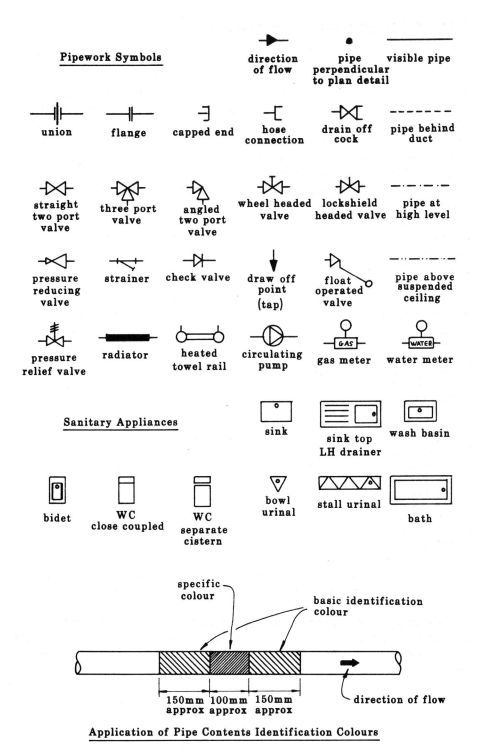

Pipework Symbols

direction of flow   pipe perpendicular to plan detail   visible pipe

union   flange   capped end   hose connection   drain off cock   pipe behind duct

straight two port valve   three port valve   angled two port valve   wheel headed valve   lockshield headed valve   pipe at high level

pressure reducing valve   strainer   check valve   draw off point (tap)   float operated valve   pipe above suspended ceiling

pressure relief valve   radiator   heated towel rail   circulating pump   gas meter   water meter

Sanitary Appliances

sink   sink top LH drainer   wash basin

bidet   WC close coupled   WC separate cistern   bowl urinal   stall urinal   bath

specific colour

basic identification colour

150mm approx   100mm approx   150mm approx

direction of flow

Application of Pipe Contents Identification Colours

Identification of Pipework

(Reproduced with permission from BS 1192: Part 3 *Recommendations for symbols and other graphic conventions* and BS 1710 *Recommendations for treatment of water for marine boilers.*)

The International Metric System

Relevant British Standard
BS 5555

The metric system was first introduced by the French National Assembly late in the six-teenth century; it was adopted by the British early in the 1970s. One of the main char-acteristics of the system is its decimal nature; the conversion from smaller to larger units and vice versa is thus made by moving the decimal point to the left or right.

The SI system of units (Système International d'Unités), developed from the metric system, has been defined and recommended as the system of choice for scientific use worldwide, used internationally by most nations. It should be noted that not all metric units are SI units, for example hectare and litre.

Examples of SI-derived units and British imperial equivalents

Quantity	Metric		Imperial		Conversion factor
	Unit	Symbol	Unit	Symbol	
Length	metre	m	yard	yd	0.9144
Area	square metre	m^2	square yard	yd^2	0.8361
Area	hectare	ha	acre	–	0.404686
Volume	cubic metre	m^3	cubic yard	yd^3	0.7646
Capacity	litre	l	gallon	gal	4.536
Mass/weight	kilogram	kg	pound	lb	0.4536
Force	newton	N	pound force	lbf	4.448
Pressure	newton/sq metre	N/m^2	pound force per square inch	lbf/in^2	6894
Pressure	Pascal	Pa		lbf/in^2	6894
Pressure	Bar	bar		lbf/in^2	0.06894
Velocity	metres per second	m/s	foot per second	ft/s	0.3048
Temperature	kelvin	K	centigrade	C	see opposite
Temperature	celsius	C	centigrade	C	see opposite
Heat energy	joule	J	British thermal unit (btu)	btu	1055
Power	kilowatt	kW	btu per hour	btuh	0.0002931
Power	kilowatt	kW	horse power	hp	0.7457

To convert metric to imperial divide by conversion factor
To convert imperial to metric multiply by conversion factor
Note: 1 Pa = 1 N/m^2;   1 bar = 100 kN/m^2

Examples of converting metric to imperial units:
  (1)   Express 15 m^2 in yd^2: 15 ÷ 0.8361 = 18 yd^2
  (2)   Express 17 kW in btu/h: 17 ÷ 0.0002931 = 58 000 btu/h

Examples of converting imperial to metric units:
  (3)   Express 50 gal in l: 50 × 4.536 = 227 l
  (4)   Express 14.7 lb f/in^2 in N/m^2: 14.7 × 6894 = 101 342 N/m^2

The metric system is designed in such a way that, by prefixing the derived unit with one of the symbols identified in the following table, its value can be increased or de-creased in multiple units, e.g. kilograms, centimetres.

Prefix	Symbol	Multiplication factor
atto-	a	10^{-18}
femto-	f	10^{-15}
pico-	p	10^{-12}
nano-	n	10^{-9}
micro-	mu	10^{-6}
milli-	m	10^{-3}
centi-	c	10^{-2}
deci-	d	10^{-1}
No prefix (SI unit only, e.g. metre)		
deca-	da	10^{1}
hecto-	h	10^{2}
kilo-	k	10^{3}
mega-	M	10^{6}
gigo-	G	10^{9}
tera-	T	10^{12}
peta-	P	10^{15}
exa-	E	10^{18}

If we put the prefix system to use we find such examples as the following expressions of derived units:

(1) 2 mm = (2 × 0.001) 0.002 m   (2) 4 cm = (4 × 0.01) 0.04 m
(3) 7 km = (7 × 1000) 7000 m   (4) 4 Mm = (4 × 1 000 000) 4 000 000 m

Converting temperature scales Three temperature scales are found in general use: the Fahrenheit, the Celsius (centigrade) and the Kelvin (absolute) scales. The Kelvin scale is measured in degrees above absolute zero; this is a hypothetical temperature, the lowest achievable, and is characterized by the complete absence of heat energy. 0°K = −273.15°C; for every 1°K rise in temperature a 1°C rise in temperature is also experienced; for example, 0°C = 273°K and 1°C = 274°K. Converting to and from the old Fahrenheit scale is achieved by using the following calculations:

Celsius to Fahrenheit =   °C × 1.8 + 32 = °F
Fahrenheit to Celsius =   (°F − 32) × 0.56 =°C

Examples: 21°C = 21 × 1.8 + 32   = 69.8°F (and 273 + 21 = 294°K)
   or   100°F = (100 − 32) × 0.56 = 38.1°C

Areas, Volumes and Capacities

Plumbers frequently need to make mathematical calculations in order to carry out their design tasks. Given here are the basic formulae used in these calculations.

Areas

When finding areas the answer is expressed in the unit squared, e.g. m^2 or mm^2, and it is thus distinguished from linear dimensions.

Area	Formula	Example No	Answer
Rectangle or square	L × B	1	$1.2 \times 0.7 = 0.84 \ m^2$
Triangle	L × D ÷ 2	2	$1.2 \times 0.7 \div 2 = 0.42 \ m^2$
Parallelogram	L × D	3	$1.2 \times 0.7 = 0.84 \ m^2$
Trapezium	Average L × D	4	$(3 + 1) \div 2 \times 0.7 = 1.4 \ m^2$
Circle	*πR^2	5	$3.142 \times 0.6 \times 0.6 = 1.13 \ m^2$

*$\pi = 3.142$, which is the amount of times the diameter will go around the circumference of a circle.

Volumes

To distinguish volume measurements the answer is expressed in the unit cubed, e.g. m^3 or mm^3, etc.

Volumes are simply found by first finding the area of the vessel or space and then multiplying by the height of the void. Examples are:

Volume	Formula	Example No	Answer
Room or tank	L × B × H	6	$1.2 \times 0.7 \times 0.6 = 0.504 \ m^3$
Cylinder/pipe	πR^2 × H	7	$3.142 \times 0.6 \times 0.6 \times 2.3 = 2.599 \ m^3$

Sometimes volume is expressed in *capacity* (or litres) in which case the volume is simply multiplied by 1000.

Thus in the last example, where the cylinder had a volume of 2.599 m^3, it would hold $(2.599 \times 1000) = 2599$ litres.

Since 1 litre weighs 1 kg we can further see that the cylinder would weigh 2599 kg or 2.6 tonnes.

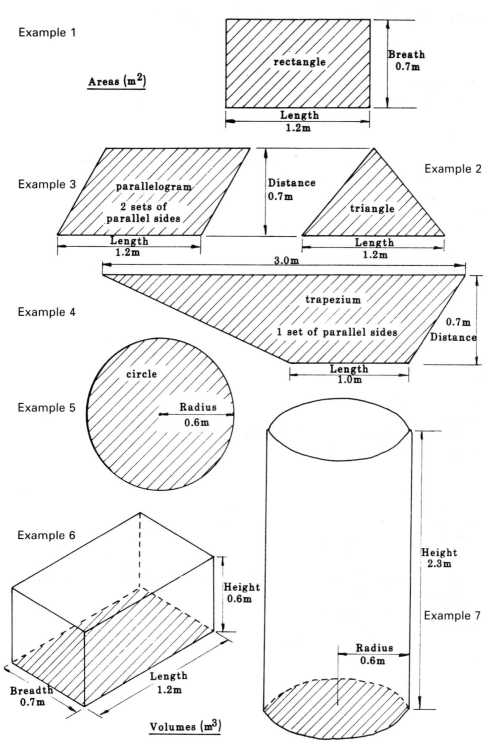

Example 1

Areas (m^2)

rectangle
Breath
0.7m
Length
1.2m

Example 3

Example 2

parallelogram
2 sets of
parallel sides
Length
1.2m

Distance
0.7m

triangle
Length
1.2m

Example 4

3.0m

trapezium
1 set of parallel sides

0.7m
Distance

Length
1.0m

circle

Example 5

Radius
0.6m

Example 6

Height
0.6m

Height
2.3m

Example 7

Radius
0.6m

Breadth
0.7m

Length
1.2m

Volumes (m^3)

Areas, Volumes and Capacities

Mass/Weight and Pressure

Relative density (specific gravity)

The weight of a substance per volume, compared to an equal volume of water, for example, 1 m³ of water weighs 1000 kg at 4°C whereas 1 m³ of lead weighs about 11 300 kg; thus it can be seen that lead is 11.3 times heavier. From this we can calculate the weight of any substance by dividing the density of a substance by the density of water. Water is always shown with a specific gravity number of 1.0; therefore any material with a number higher than 1.0 will sink in water and those with a number less than 1.0 will float.

The relative density of gases can also be measured and are compared with the specific gravity of air which is also expressed as 1.0.

It is worth noting that we can compare materials with each other; for example, we can see that in fact cast iron weighs less than tin per equal volume and natural gas is lighter than air and will rise upwards should there be a leak.

Relative density at atmospheric pressure

Material	Relative density
Solids and liquids	
Water	1.0
Class C fuel oil	0.79
Linseed oil	0.95
Aluminium	2.7
Zinc	7.1
Cast iron	7.2
Tin	7.3
Mild steel	7.7
Copper	8.9
Lead (milled)	11.3
Mercury	13.6
Gases	
Air	1.0
Methane (natural gas)	0.6
Propane	1.5
Butane	2.0

Density

The number of molecules per volume in a particular substance. As previously stated, 1 m³ of water at 4°C weighs 1000 kg whereas a cubic metre of water at 82°C weighs only 967 kg; thus it can be seen that the water at the lower temperature is heavier, there being more molecules, and the molecules are closer or 'more dense' (more compact).

Water, when cooled below 4° Celsius, begins to expand again; this is a strange phenomenon peculiar to water. It will be seen that water is at its maximum density at 4°C.

Pressure and forces

Pressure can be defined as the force acting upon a given area. Force on the other hand can be defined as any action that tends to hold or alter the position of a body. Since the introduction of the metric system in this country the unit of force has been the newton (N).

One newton is the force that gives a mass of 1 kg an acceleration of 1 metre per second squared (1 N = 1 kg × 1 m/s²).

The force produced by a 1 kg mass = 9.81 N

Sometimes other terms are used to identify pressure such as the pascal, the bar or the pound force per square inch. These can be illustrated as follows:

$$1\ Pa = 1\ N/m^2$$
$$1\ bar = 100\ 000\ N/m^2\ (100\ kN/m^2)$$
$$1\ lbf/in^2 = 6894\ N/m^2\ (6.894\ kN/m^2)$$

Atmospheric pressure

The pressure created by the weight of the atmosphere pushing down on the Earth. The pressure at the top of a mountain is different from that in a valley below sea level. The pressure created at sea level would be 101.3 kilonewtons per square metre (101.3 kN/m²).

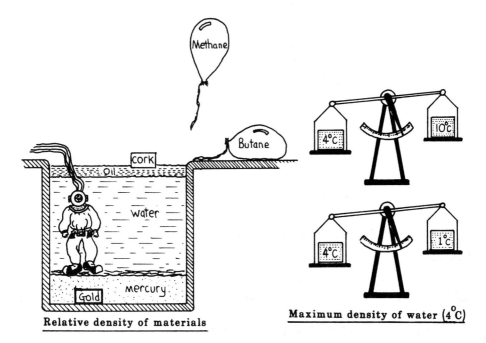

Relative density of materials

Maximum density of water (4°C)

Mass/Weight and Pressure

Water Pressure

There are two basic means of creating pressure in any plumbing system: (1) by means of a pump connected to a pipeline; or (2) by means of the weight of the water itself. The higher a column or head of water, the more pressure is exerted at its lowest point; therefore it is essential to install feed and storage cisterns as high as possible, giving a good pressure at the draw-off points. A container 1 m × 1 m × 1 m would contain 1000 litres of water and weigh 1000 kg; therefore the 1000 kg of water would exert $1000 \times 9.81 = 9810$ newtons per square metre (9.81 kN/m²). If a container 1 m high exerts a pressure of 9.81 kN/m², a container 4 m high simply exerts four times as much pressure.

Example: Find the pressure exerted at the base of a container 4 m high with a base 1 m long and 1 m wide. Answer: $4 \times 9.81 = 39.24$ kN/m²

The pressure found so far is the total pressure acting upon an area 1 m by 1 m and is known as the intensity of pressure.

$$\text{Intensity of pressure} = \frac{\text{force}}{\text{area}} = \frac{\text{kN}}{\text{m}^2}$$

Should the total pressure need to be found for an area larger or smaller than 1 m² then first the intensity of pressure is found and it is simply multiplied by the area which is being acted upon.

Example: Find the total pressure at the base of a boiler which measures 500 mm long, by 300 mm wide, fitted 2.5 m below the water level in a storage cistern.

$$
\begin{aligned}
\text{Total pressure} &= (\text{intensity of pressure}) \times (\text{area acted upon}) \\
&= (\text{head} \times 9.81\ ) \times (\text{length} \times \text{width}) \\
&= 2.5 \times 9.81 \times 0.5 \times 0.3 = 3.6788\ \text{kN}
\end{aligned}
$$

By transposition of the formula the height to which the water would rise in a water main can be found.

Example: The mains water pressure is 400 kN/m² (4 bar). Ignoring any frictional resistances find the height to which the water will rise in a vertical pipe.

$$\text{Intensity of pressure} = \text{head} \times 9.81$$

Transposition of formula:

$$\text{Head} = \frac{\text{Intensity of pressure}}{9.81}$$

$$\text{Therefore} \quad \frac{400\ \text{kN/m}^2}{9.81} = 40.77\ \text{m}$$

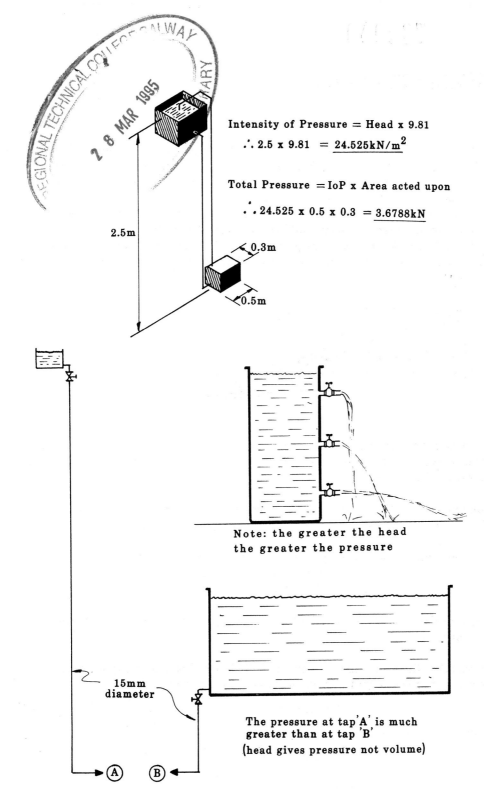

Intensity of Pressure = Head x 9.81

$\therefore$ 2.5 x 9.81 = $\underline{24.525 \text{kN/m}^2}$

Total Pressure = IoP x Area acted upon

$\therefore$ 24.525 x 0.5 x 0.3 = $\underline{3.6788\text{kN}}$

2.5m

0.3m

0.5m

Note: the greater the head
the greater the pressure

15mm
diameter

The pressure at tap 'A' is much
greater than at tap 'B'
(head gives pressure not volume)

(A)   (B)

Water Pressure

Physical Properties of Materials

The term *physical properties* refers to the weight, colour and strength, etc., of a material. All materials have their own qualities and properties, the meaning of each being identified as follows:

Density The mass of molecules per volume in a particular substance, thus its weight, e.g. kg/m^3 (see relative density).

Ductility The ability to withstand distortion without fracture, e.g. a metal drawn out into a fine wire.

Durability The ability to resist wear and tear, i.e. long lasting.

Elasticity The ability of a metal to return to its original shape after being distorted.

Fusibility The melting point of the material, i.e. when a solid changes to a liquid.

Hardness The ability of a material to resist penetration or wear. The hardest natural substances to be found are diamonds.

Malleability The ability of a metal to be worked without fracture.

Plasticity The exact opposite of elasticity. A material which does not return to its original shape when deformed.

Relative density (also called specific gravity) The weight of a substance compared to an equal volume of water. Water has a relative density of 1; materials with a number greater than 1 will sink in water.

Specific heat The amount of heat required to raise 1 kg of material 1°C.

Temper The degree of hardness in a metal.

Tenacity A material's ability to resist being pulled apart.

Tensile strength A measure of tenacity contained in a material (measured in N/m^2).

Thermal expansion The amount a material expands when heated.

Categories of metals

Metals can be divided into two groups, either ferrous or non-ferrous. Ferrous metals are those which contain iron; conversely, non-ferrous metals do not. The main difference between these two categories is that, with the exception of stainless steel, ferrous metals will rust on exposure to water and oxygen. Also ferrous metals can be magnetised.

Physical properties of common plumbing metals

Material	Chemical symbol	Melting point (°C)	Boiling point (°C)	Density (kg/m³)	Specific gravity	Specific heat (kJ/kg°C)	Tensile strength (MN/m²)
Aluminium	Al	660	2467	2705	2.7	0.887	81
Copper	Cu	1083	2529	8900	8.9	0.385	308
Iron	Fe	1535	3000	7860	7.8	0.554	278
Lead	Pb	327	1744	11300	11.3	0.125	16
Tin	Sn	232	2260	7310	7.3	0.226	23
Zinc	Zn	419	907	7130	7.1	0.397	170

Metals can be further divided into either pure or element metals and alloys. Pure metals are those which do not have any other metal or material mixed with them; examples can be seen in the previous table. An alloy on the other hand is a mixture of two or more metals or a metal mixed with a non-metal.

Some common alloys

Alloy	Main elements
Brass	Zinc and copper
Bronze	Copper and tin
Cast iron	Iron and 2–4% carbon
Gunmetal	Copper, tin and zinc
Mild steel	Iron and approximately 0.2% carbon
Stainless steel	Iron, chromium and nickel
Solder	Lead and tin
Solder (lead free)	Tin and copper

Categories of plastics

Plastics are man-made materials and can be divided into two broad groups: thermo-plastics and thermosetting plastics. A thermoplastic, when heated, will soften; in this state it can be formed to the shape required and on cooling will harden. If required it can be reheated to form a new shape, without any significant change in properties. A thermosetting plastic on the other hand can be heated initially to soften and form the shape, but when it cools the shape is permanent and no reheating will soften it again.

- ☐ **Thermoplastics** These include acrylics (Perspex), nylon, polythene, polypropylene and polyvinyl chloride (PVC).
- ☐ **Thermosetting plastics** These include Bakelite, Formica, melamine and polyester.

Heat and its Effects

Quantity of heat

The amount of heat contained in a substance and its temperature are two completely different things. A burning match has a much higher temperature than an iceberg, but the total heat energy in the iceberg is much greater than that of the match. The heat to raise 500 litres of water to 65°C is greater than that required to raise 5 litres of water to 65°C; the temperatures may be the same but the 500 litres of water contain much more heat and equally give off much more heat when cooling. The quantity of heat a substance contains is measured in joules (J)

1 joule = the power supplied by 1 watt in 1 second.

To increase the temperature of 1 g of water by 1°C, 4.186 J would be required. Therefore to heat 1 litre or 1 kg (1000 g) of water 1°C 1000 × 4.186 J = 4186 joules or 4.186 kJ would be required.

Specific heat

Specific heat is the amount of heat required to raise 1 kg of material 1°C. The heat required would differ from material to material, e.g. while it would require 4.186 kJ to raise the temperature of water by 1°C, only 0.385 kJ would be needed to raise the temperature of copper by 1°C.

It must be noted that specific heat values vary as the temperature changes.

Specific heat values

Material	kJ/kg°C
Water	4.186
Aluminium	0.887
Cast iron	0.554
Zinc	0.397
Lead	0.125
Copper	0.385
Mercury	0.125

Transference of heat

Heat can be transferred in the following ways:

Conduction: Heat passing through or along a solid.

Convection: The molecules of liquids and gases expand in size; they thus become lighter in volume (i.e. fewer molecules can occupy the same space), and rise, being pushed upwards by the cooler, more dense liquid.

Radiation: The direct heat felt from a heat source (e.g. the sun).

Calorific value

Calorific value basically means the amount of heat produced by a specific amount of fuel when completely consumed. The calorific value of fuels differs from one fuel to another;

for example, the amount of heat produced by burning 1 kg of wood is less than that produced by burning 1 kg of coal. Natural gas has an even higher calorific value.

The calorific value of gases is usually expressed as the amount of heat in joules they contain per cubic metre (m³), while for solid and liquid fuels it is expressed in joules per kilogram.

Some calorific values

Fuel	Calorific value
Anthracite (coal)	32 MJ/kg
Wood	19 MJ/kg
Domestic grade oil	45 MJ/kg
Liquefied petroleum gas	30 MJ/kg
Natural gas	38 MJ/m³
Town gas	19 MJ/m³
Electricity	3.6 MJ/kW

Thermal expansion

Most materials expand when heated. All substances are made up of molecules which, when heated, move about more vigorously and thus move further apart, which results in the material becoming larger. When the material cools the molecules slow down and move closer together; thus the material gets smaller or contracts. The amount the material expands in length when heated can be measured by a simple calculation using the following formula:

length × temperature rise × coefficient.

The coefficients of thermal expansion for typical plumbing materials are given in the following table:

Material	Coefficient °C
Plastic	0.00018
Zinc	0.000029
Lead	0.000029
Aluminium	0.000026
Tin	0.000021
Copper	0.000016
Cast iron	0.000011
Mild steel	0.000011
Invar	0.0000009

Example: Find the amount a 9 m long plastic discharge stack will expand due to a temperature rise of 24°C.

$$9 \times 24 \times 0.00018 = 0.039 \text{ m or } 39 \text{ mm}$$

Physical change

A physical change happens when a substance changes from a liquid to a solid or a liquid to a gas. For example, water (liquid) changes to ice(solid) by cooling to 0°C or to steam (gas) by heating to 100°C. In all three states – solid, liquid or gas – it remains H_2O, and the change made is only temporary due to outside forces such as heat or pressure. As another example, lead, if heated to 327°C, melts, and if further heated to 1740°C it changes to a lead vapour.

Corrosion

The destruction of a metal resulting from a chemical reaction on its surface. Several types of corrosion can be identified, the most common being atmospheric and electrolytic corrosion and dezincification of brass.

Atmospheric corrosion (oxidation)

Corrosion caused by moisture and gases in the air. When oxygen mixes with the surface of the metal it forms an iron oxide, commonly called rust. This rust falls away, exposing fresh metal underneath, and the process continues until the metal rusts away completely. Non-ferrous metals are attacked by such gases as carbon dioxide and sulphur dioxide, but surface corrosion which occurs does not flake off like rust, and therefore acts as a skin on the surface (called patina) and protects the metal from further corrosion. The colour of patina is different from that of the roofing metal itself, the most striking example being copper, which turns green. In coastal areas there is a lot of salt in the air; the rain would be a strong alkali in these areas, and would tend to destroy aluminium.

Electrolytic corrosion (galvanic action or electrolysis)

The destruction of one metal (the anode) by another (the cathode) when connected together via an electrolyte (any liquid or moisture which carries electrically charged particles from an anode to a cathode), such as water. The rate of corrosion depends upon the water and the distance apart the metals are on a list of metallic elements known as the electromotive series. If the water is acidic or hot, the corrosion will be increased. The following elements are to be found in the electromotive series:

□ Copper
□ Lead
□ Tin
□ Nickel
□ Iron
□ Zinc
□ Aluminium
□ Magnesium

Note: Those high on the list will destroy those lower on the list, e.g. copper will destroy zinc.

Dezincification of brass

A condition where the zinc in brass has been destroyed, leaving the fitting porous and brittle. The zinc content of the alloy is converted into a basic carbonate, which also causes a blockage in the pipeline. Basically dezincification is a form of accelerated electrolytic corrosion (brass being composed of copper and zinc).

Occurrence of corrosion

To prevent electrolytic corrosion on sheet roofwork one must avoid the use of mixed metals and never let metal roofs higher on the electromotive series discharge onto roofs lower on the list, e.g. lead onto aluminium will cause excessive corrosion.

Corrosion within plumbing systems occurs due to several design faults. To prevent atmospheric corrosion never use 'black iron' pipes (i.e. not galvanised) if the water is to pass through the system, such as on mains supply or hot and cold distribution pipework, as air had been absorbed into the water when it fell as rain.

Black iron pipes and steel radiators can be used in heating systems without fear of atmospheric corrosion because, providing the system is designed correctly, once the water has filled the heating pipework, it simply circulates round the system and is never replaced. Thus, as the air is expelled from the water, via its circulation, corrosion cannot occur.

Corrosion of metals can be caused by materials such as new timbers made of red cedar, chestnut, fir, oak and teak, which can leak harmful acids, as does the growth of moss and lichen on roofs. Alkaline attack can occur due to the presence of fresh cement, lime and plaster.

Prevention of corrosion

Protective coatings The simplest way of protecting metals is to paint or cover them with plastic, etc. Iron is often dipped in zinc (a process known as galvanising) to give it a zinc coat, this prevents the air touching the surface of the metal and causing attack.

Cathodic protection To prevent electrolytic corrosion a metal is often incorporated within the system which will be destroyed before all others; this is known as the sacrificial anode. Examples include galvanising or the rod of magnesium which is fixed into many hot storage vessels. Sometimes a piece of aluminium is placed into the cold storage cistern to give the same effect.

System design To reduce corrosion problems the system needs to be designed to prevent air entering the pipework. The most important consideration in vented hot water heating system design is that the water is filled via a feed and expansion cistern (f & e) separate from that for the domestic water distribution pipework to the sanitary appliances; otherwise the waters will mix in the storage cistern upon expansion.

Corrosion inhibitors These are used as a method of preventing corrosion in heating systems. A chemical is added to the primary heating circuit via the f & e cistern. This neutralises any flux residues, etc., which make the water acidic; it also coats the system with a fine film.

Specialist Hand Tools for Pipework

Pipe cutters

Apart from hacksaws, which obviously could be used, pipe cutters fall into three basic types: roller, wheel and link cutters.

The **roller cutter** is the most commonly used, especially for smaller pipes such as copper tube; in fact the 'pipe slice' is now a very familiar tool, enabling easy access in awkward locations. The roller cutter consists of two rollers and one cutting wheel. The pipe cut is achieved by rotating the cutter fully around the pipe. The rollers prevent a burr from forming on the external pipe wall. An internal burr will, however, be created in using these cutters, and this must be removed, using either a round file or a deburring reamer.

The **wheel cutter** consists of three to four cutting wheels and is suitable for cutting pipe where a full 360° turning circle cannot be achieved; it does, however, have the drawback of leaving an additional external burr on the pipe.

Link cutters are used on large diameter cast iron pipes. When cutting cast iron no burr is produced because the pipe actually breaks owing to the even pressures achieved and the brittle nature of the pipe wall.

Pipe wrenches and spanners

Among the most important of these tools used are straight pipe wrenches, pipe grips, chain wrenches, basin spanners and adjustable spanners. The wrench has teeth which are used to bite into the pipe. The spanner has smooth-faced jaws which are designed to be used across the flats of nuts, thus preventing the damage and slipping which may occur using a wrench.

Pipe threading equipment

Sometimes a thread needs to be cut on steel pipework; this is achieved using dies. The dies are held in a die stock which can be of several designs and be either hand held or machine operated. The hand held design is commonly of a drop head stock in which a separate die head is used for each size of pipe; this is not so bulky as those requiring adjustment of the thread size.

Pipe bending equipment

Many designs of pipe bender will be found. In general those used for copper pipe are of the lever design, the type used for smaller pipe sizes having fixed rollers. Where roller adjustment is required incorrect pressure will have one of two effects: if the adjustment is too low it will be too tight and will cause excessive throating (flattening) to the finished bend; if it is too high it will be too loose and ripples on the pipe will be produced. When bending copper tube, full support of the pipe is required, which is achieved using a backguide. For mild steel pipes a hydraulic bending machine will be used, with no backguide support necessary.

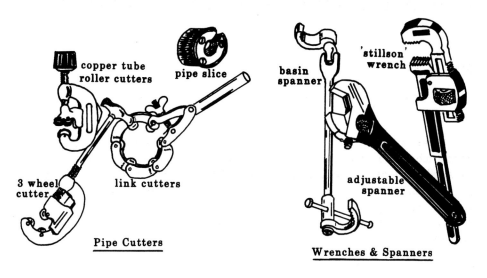

copper tube
roller cutters

pipe slice

basin
spanner

'stillson'
wrench

3 wheel
cutter

link cutters

adjustable
spanner

Pipe Cutters

Wrenches & Spanners

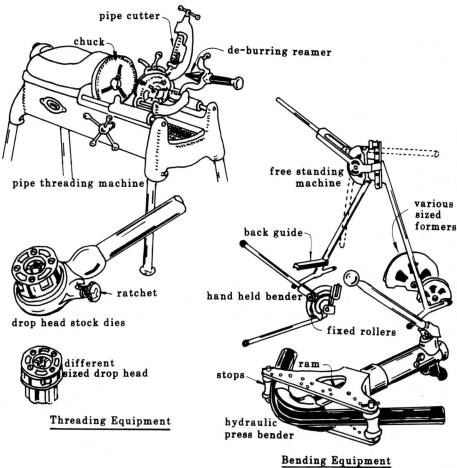

pipe cutter

chuck

de-burring reamer

free standing
machine

various
sized
formers

pipe threading machine

back guide

ratchet

hand held bender

fixed rollers

drop head stock dies

ram

stops

different
sized drop head

hydraulic
press bender

Threading Equipment

Bending Equipment

Specialist Hand Tools for Pipework

Specialist Hand Tools for Sheetwork

Many tools are used when carrying out the weathering to a roof using lead and copper sheet, etc. Among these are tin snips, mallets and hammers to name a few; but the plumber also uses specialist tools which are made from suitable hardwoods such as box, beech or hornbeam, or to avoid high costs, materials such as high density polythene. Typical tools include the following:

Flat dressers: used to dress the metal to a flat surface, making it lie flat without any undue humps.

Bossing tools: used to boss (form) sheet lead into the required shape. The tool used to boss the lead will be to the preference of the user and includes the bossing mallet, bossing stick or bending stick. Today many plumbers use a rubber mallet.

Setting-in stick: a tool used to set in a crease in sheet lead or reinforce and square up the metal on completion. The tool is struck with a mallet.

Chase wedge: used as a setting-in stick or to chase in angles and drips when used in conjunction with a drip plate.

Step turner: used to turn in the individual steps on step flashings to allow fixing into the brickwork.

Drip plate: a piece of steel plate, 100 mm × 150 mm × 1 mm thick, which is positioned between two pieces of lead being bossed close together; it allows one sheet to slide freely over the other.

Dummy: a home-made tool designed to assist in starting the corner when lead bossing; it is also used to assist bossing difficult details and angles.

Lead knife: a special knife which is used to cut sheet lead. The lead is scored with the knife several times and pulled apart to give a neat straight cut.

Shavehook: used to remove the thin layer of oxide on sheet lead prior to lead welding.

Seaming pliers: these tools are only used for the harder sheet materials such as aluminium, copper and zinc and are used to assist in folding and forming welts.

To give a long life to wooden tools they should regularly be soaked in linseed oil and never struck or stored with steel tools. The edges should be kept slightly rounded to prevent damage to the sheet metal being worked.

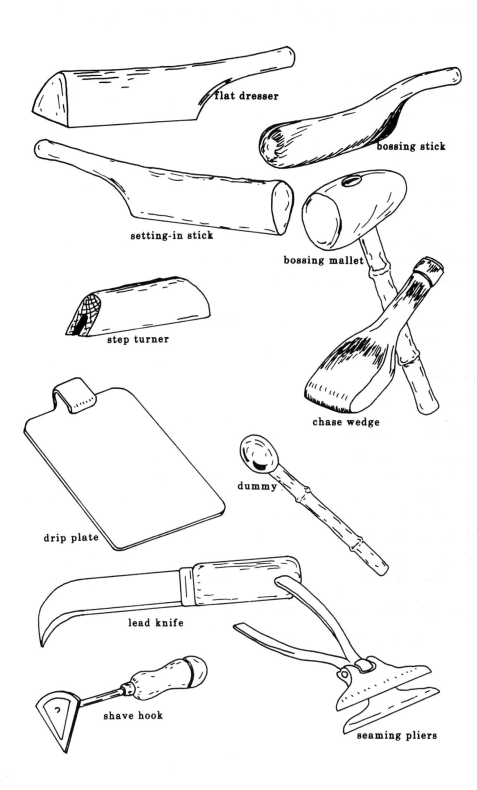

flat dresser

bossing stick

setting-in stick

bossing mallet

step turner

chase wedge

drip plate

dummy

lead knife

shave hook

seaming pliers

Specialist Hand Tools for Sheetwork

Plastic Pipe and Fittings

Relevant British Standards
BS 4514, BS 5255 and BS 4346

Two common types of plastic will be found in the construction industry. Firstly there are those produced by the polymerisation of ethylene; these include polythene, polyethylene and polypropylene, which are mainly used for mains supply pipework, hot water and heating systems and certain waste pipe applications. The second type include uPVC (uplasticised polyvinyl chloride) and ABS (acrylonitrile butadine styrene) which are used primarily for waste distribution pipework and cold water installations.

Plastic can be joined by either compression fittings, push fit connections, fusion welded joints or solvent welded joints; the method chosen will depend on the type of plastic and its use. For example, it is possible to form a solvent weld joint to uPVC, but it cannot be fusion welded; conversely polythene cannot be solvent welded.

Compression fittings

These are of two types: (1) those used for polythene, made in gunmetal or similar, and used on supply pipework to withstand high pressures. These use a copper compression ring with a copper liner, which is inserted into the pipe to prevent the wall collapsing when the nut is tightened; (2) those (as used on uPVC) usually made of a plastic material and used for waste discharge pipework; this type utilises a rubber compression ring, the fitting only being made hand tight.

Push fit connections

Many types of push fit joint are now being marketed for both high pressure pipework and waste systems. The main difference between them is that those used on high pressures incorporate a grip ring which prevents the pipe from pulling out. Those used for waste pipework simply have a rubber 'O' ring or the equivalent. Sometimes a push fit connection is used on waste pipework to allow for expansion and contraction; if this is the case when first assembling the pipe fitting push the pipe in fully and then withdraw it by about 10–15 mm.

Fusion welded joints

This is a joint in which the plastic is melted onto the fitting. Materials such as polythene and polypropylene are joined by this method. The join is achieved by the use of a specially heated tool which melts the pipe and fitting; or an electric charge is applied to a wire which is located just below the surface plastic of special fittings. This wire heats up and melts the plastic.

Solvent welded joint

This joint, made from a special solvent cement, is used to join materials such as uPVC and ABS. Solvent cement is not a glue; it does not simply stick the pipes together but when the liquid is applied to the plastic it temporarily dissolves it. A solvent weld joint will set in 5–10 minutes, but will require at least 12–24 hours to become fully hardened.

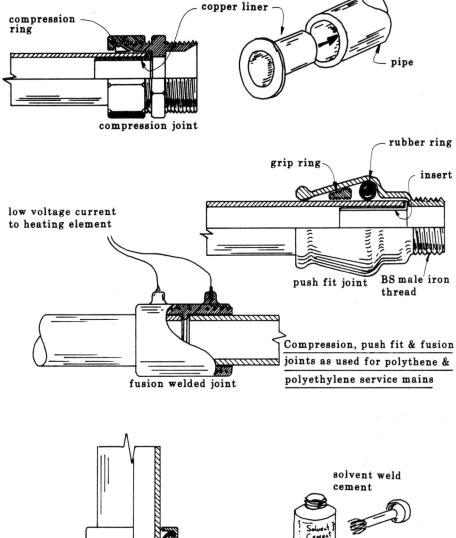

compression ring

copper liner

pipe

compression joint

rubber ring

grip ring

insert

low voltage current to heating element

push fit joint

BS male iron thread

fusion welded joint

Compression, push fit & fusion joints as used for polythene & polyethylene service mains

solvent weld cement

Solvent Cement

push fit joint

allowance for expansion

uPVC pipe

solvent welded joint

Solvent welded and push fit joints used for low pressure plastic pipework

Plastic Pipe and Fittings

Low Carbon Steel Pipe and Fittings

Relevant British Standard
BS 1387

Low carbon steel pipe, often referred to as mild steel, is available either painted black or galvanised (coated inside and out with zinc). Black iron pipes should only be used for wet heating systems or oil and gas supply pipework. If it were used where fresh water is continuously being drawn off through the pipeline it would soon be attacked by corrosion. Steel pipes are manufactured in three grades and colour-coded accordingly: heavy gauge, red; medium gauge, blue; and light gauge, brown. The outside diameter in each case is the same. Only heavy gauge is permitted to be used below ground. All grades are used above ground although light gauge tube is rarely employed, being restricted, possibly, to dry pipe sprinkler systems.

Methods of jointing include threaded joints, compression joints and welded. Welding is restricted to black iron pipework only; welding to galvanised iron pipes would result in the zinc being burnt off.

Threaded joints

Threads are cut into the pipe to give a BS pipe thread (BSPT) to BS 21 using stocks and dies. The thread is cut on site using hand dies or powered threading machines. An assortment of fittings is used to join the pipe which may have either a parallel or tapered thread and may be made from either malleable cast iron or steel. Steel fittings tend to be stronger, although more expensive.

The length of the thread on the pipe should be such that one and a half or two threads are showing when the fitting is assembled. This is because these first two threads have not been cut to the correct depth, owing to the cutting process. A few strands of hemp may be applied to the thread in a clockwise direction followed by some jointing paste to give a sound joint. An alternative jointing material is polytetrafluoroethylene tape (PTFE) which can be used, but should be avoided on larger threads or where the thread is slightly distorted. The jointing medium used will depend on the contents of the pipe.

Compression joints

It is possible to purchase several designs of compression joint for use on steel pipe; these incorporate a rubber compression ring They tend to be rather expensive but a saving can be made on installation time.

Disconnection joints

Because of the way in which screwed fittings are made, i.e. rotated on the pipe, it is impossible to remove or assemble pipework where this rotation is not possible. Therefore a special fitting will be required such as that provided by a union, long screw or a flange, which allow for the joint to be made and disconnected without any turning of the pipe or fitting.

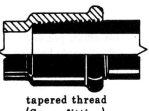

tapered thread
(Crane fitting)

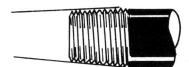

male thread cut onto pipe
with stocks & dies

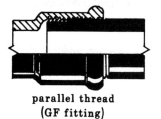

parallel thread
(GF fitting)

female thread of fittings
as produced by two leading
manufacturers

stainless steel
backing washer

rubber
compression ring

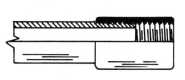

threaded socket being used

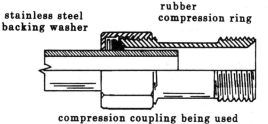

compression coupling being used

correct order of tightening
bolts on a flange to
prevent distortion

gasket

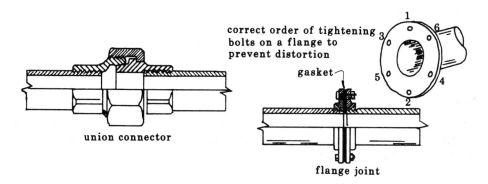

union connector

flange joint

longscrew

Low Carbon Steel Pipe and Fittings

Copper Pipe and Fittings

Relevant British Standards
BS 2871, BS 2051 and BS 864

Copper pipe is available in the following grades: W, X, Y and Z. The basic difference between the grades of pipe is the wall thickness and temper. The outside diameter of the pipe always remains the same.

Grade X is by far the most common and used in general purpose work.
Grade Y is softer and thicker walled, and is usually supplied in coils.
Grade Z is hard tempered with thin walls and is unbendable.
Grade W is used for micro-bore heating systems.

Copper pipe is jointed by several methods including compression fittings, soldered joints and push fit connections.

Compression joints

There are two main types: manipulative and non-manipulative. The difference is that with the manipulative joint the end of the pipe has to be worked to form a bead or bell mouth (i.e. manipulated) whereas with the non-manipulative joint the nut is simply slipped on followed by the compression ring (olive). It must be noted that joints used below ground must be of the manipulative type.

Soldered joints

A soldered joint is one in which alloys with a lower melting temperature than the copper are used to join the pipe. There are two basic classifications of soldered joint: soft- and hard-soldered.

Soft-soldered joints are made with a propane blowtorch, using tin composite solders melting in the range 180–230°C. The solder is drawn into the fitting by capillary attraction. Hard soldering requires much higher temperatures – in the region of 600–850°C. These are usually achieved using oxyacetylene equipment.

A hard-soldered joint can be formed in much the same way as a soft-soldered joint, or the pipe can be manipulated to form a bell mouth opening in which large volumes of solder are deposited to form what is known as a bronze welded joint. It is possible to purchase soft-soldered joints with a ring of solder already in, thus ensuring that the correct amount of solder is used; alternatively the solder may be added.

Since the introduction of the 1989 local water byelaws solders containing lead are not permitted to be used for pipework which serves hot and cold drinking water supplies.

Push fit joints

There are several types of push fit joint designed for use on hot and cold water supply up to 1.5 bar pressure; they are generally made of plastic with a neoprene 'O' ring. The biggest problem with these joints is their bulky and unsightly appearance and the problem of maintaining the electrical continuity of the house equipotential bonding.

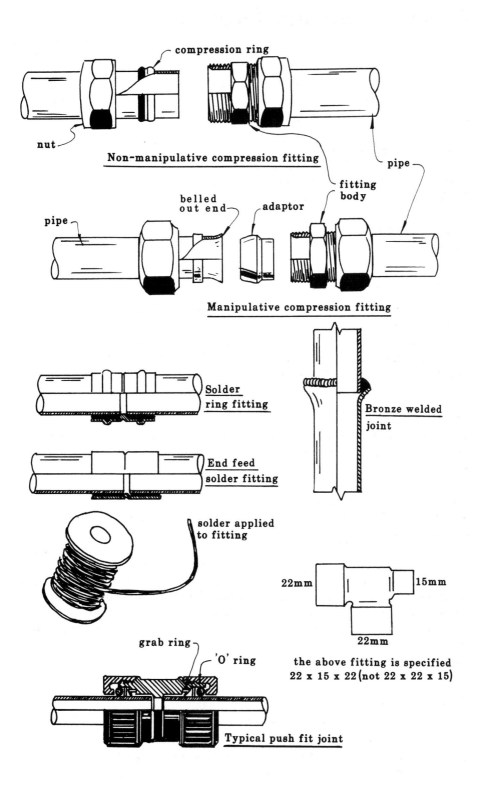

compression ring

nut

Non-manipulative compression fitting

pipe

fitting body

belled out end

adaptor

pipe

Manipulative compression fitting

Solder ring fitting

End feed solder fitting

Bronze welded joint

solder applied to fitting

22mm

15mm

22mm

the above fitting is specified 22 x 15 x 22 (not 22 x 22 x 15)

grab ring

'O' ring

Typical push fit joint

Copper Pipe and Fittings

Low Carbon Steel Pipe Bending

Steel pipes are bent either by using a hydraulic bending machine or by applying heat to the section to be bent. Due to the thickness of the pipe walls it is not necessary to support the pipe fully, as with bending copper tube.

Hydraulic press bender

To use a hydraulic press bending machine a former of the correct size is selected and fitted onto the end of the ram. The stops are positioned into the pin holes as indicated on the machine. When the pipe is inserted into the former and the handle operated the ram moves forwards to force the pipe against the stops; as the pumping continues the pipe is forced to bend. To release the pumping action a by-pass valve is turned which allows the hydraulic fluid to escape to another chamber in the machine.

Bending using the hydraulic press bender It is very simple to form a 90° bend from a fixed point on these machines. First a line is marked on the pipe at a distance from the fixed point to that equal to the fixed point and the centre line of the finished bend (distance 'X', see figure). Then, from this measurement is deducted the internal bore of the pipe. The pipe is now placed in the bending machine at this new point and lined up with the centre line of the correct sized former. The machine is now pumped to apply pressure and bend the pipe to the required angle. Due to a certain amount of elasticity the bend is over-pulled by about 5°.

To make an offset using these machines first measure along the pipe from the fixed point (this time making no deductions) and pull this round to the required angle, as shown. Then, using a straight edge or positioning the pipe against two parallel lines the distance of the offset, the second bending point is marked on the tube. The pipe is replaced in the machine and this mark is positioned at the centre line of the former. When pulling this second bend it is essential to ensure that it is pulled running true and on the same plane as the first bend.

Bending using heat

This is described on page 42 under 'pipe bending using heat'. Some of the main points to observe when using heat to bend steel tube include:

☐ Mark out the heat length correctly and apply heat, until cherry red, along the whole heat length;
☐ Pull as much as possible in one go, pulling from each end alternately;
☐ Should a second heat-up be required, reheat the whole heat length;
☐ As a point of safety, always cool completed work.

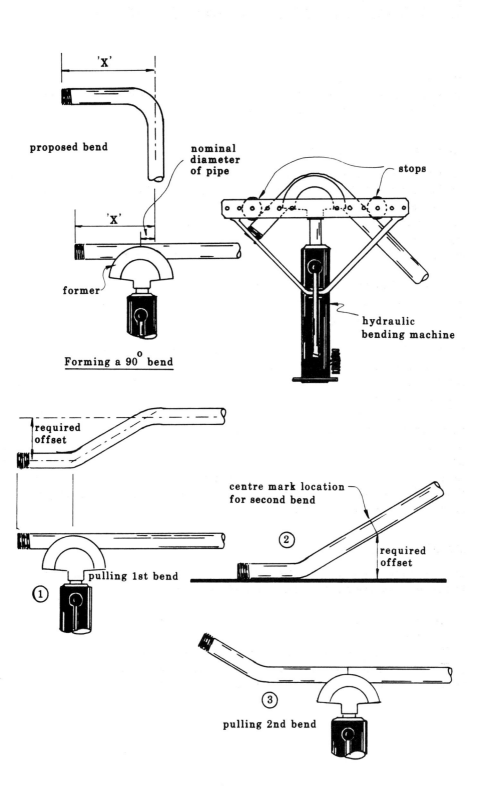

'X'

proposed bend

nominal
diameter
of pipe

stops

'X'

former

hydraulic
bending machine

Forming a 90° bend

required
offset

centre mark location
for second bend

②

required
offset

pulling 1st bend

①

pulling 2nd bend

③

Low Carbon Steel Pipe Bending

Copper Pipe Bending

Hand bending

This is a method of preventing the pipe from collapsing either by using a spring or by packing the pipe with dry silver sand. The spring could be internally inserted or fitted externally to the pipe. When using an internal spring the radius of the bend must not be too small or it will be impossible to withdraw the spring. To assist spring removal slightly over-pull the bend, then pull back to the required position. Larger pipes will need the application of heat (see page 42).

Machine bending

Bending machines are used for copperwork on the principle of leverage and are either hand held or free standing. The larger machines have different sized formers and back-guides for the various pipe sizes. Once the machine is set up, with the rollers correctly adjusted to prevent throating or rippling, the various bends can be pulled as follows:

90° bend A line is marked on the pipe at a distance from the end of the pipe equal to that between the fixed point and the back of the finished bend (distance 'X', see figure). The pipe is then inserted into the bending machine as shown. Make sure the line marked on the pipe is square with the back of the former; the bend is simply pulled round by the lever arm.

An offset Initially the first bend is pulled to any angle, unless one has been specified. The pipe is then pushed further along the bending machine and turned around to make the bend just pulled look upwards away from the former. A straight edge and rule are now used, as shown, to measure the required offset; when marking out, ensure that the straight edge is running parallel with the piece of pipe looking upwards away from the former. Once the correct measurement is obtained the bend can be pulled round until it is at the same angle as the first bend and can be seen to run parallel; this is best checked using a straight edge.

The passover There are several methods of forming a passover, one of the simplest being as follows: a bend is first pulled, the angle of which depends upon the size of the obstruction; the bend should not be too sharp, otherwise difficulty will be experienced when pulling the offset bends. A straight edge is positioned over the bend centrally at a distance equal to the required passover. The pipe is marked with two lines (see figure), these are the back of the finished offsets. The pipe is then inserted into the bending machine and when the first mark is in line with the back of the former the first set is pulled; once completed, the pipe is simply reversed in the machine and the second mark is lined up in the former and pulled to complete the passover.

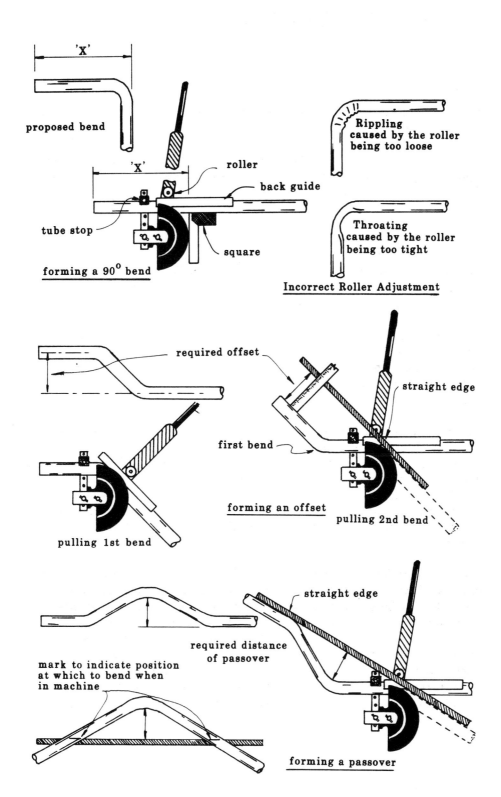

'X'

proposed bend

roller

back guide

tube stop

square

forming a 90° bend

Rippling
caused by the roller
being too loose

Throating
caused by the roller
being too tight

Incorrect Roller Adjustment

required offset

straight edge

first bend

forming an offset

pulling 2nd bend

pulling 1st bend

straight edge

required distance
of passover

mark to indicate position
at which to bend when
in machine

forming a passover

Copper Pipe Bending

Pipe Bending Using Heat

By the application of heat, it is possible to pull a bend on plastic, copper and low carbon steel pipe without the use of a bending machine. Two principles need to be understood in order to achieve a successful bend. Firstly, the internal diameter of soft- or thin-walled pipes must be protected from flattening; this is achieved by inserting a bending spring, or packing the pipe with dry silver sand. Steel is bent without any internal support. Secondly, when heating the pipe, the bend must be heated to the correct length only. The heat length is found either by calculation or by producing a drawing from which to take a measurement.

The heat length (by calculation)

The amount of pipe to be heated is simply found using the following formula:

$$2R\pi \div 4$$

$$R = \text{centre line radius of proposed bend}$$
$$\pi = 3.142$$

In the absence of a specified radius one generally uses four times the nominal diameter of the pipe.

Example: Find the heat length for a 15 mm (½ in BSP) diameter low carbon steel pipe which is to be pulled to an angle of 90°.

$$\text{Thus radius} = 15 \times 4 = 60 \text{ mm}$$
$$\text{Therefore } 2R\pi \div 4 = 2 \times 60 \times 3.142 \div 4 = 94 \text{ mm}$$

Sometimes, to save time on site and to avoid using a calculator, the plumber simply breaks down the formula to the simplified version of one and a half times the centre line radius.

Therefore, if we again find the heat length for the previous example, using this method we find that:

$$R = 15 \times 4 = 60$$
$$\text{Therefore } 60 \times 1\tfrac{1}{2} = 90 \text{ mm}$$

which is not too far from the previous example, and hence good enough.

To bend the pipe the heat length is chalked onto the pipe, and the pipe is heated as necessary and pulled 45°, quickly reversing the pipe and pulling the second 45°, if possible, to achieve the 90° bend. It is not likely to be possible to pull bends of this angle in pipes greater than 20 mm in one go.

In bending angles of less than 90° the heat length is generally taken to be half the total heat length for a 90° bend.

When bending a metallic pipe the pipe should be heated to an annealed temperature (red hot) whereas with thermoplastics the material is heated to a temperature at which it becomes soft and pliable. The softening temperature of plastic is very near the temperature at which it chars; therefore care must be taken not to damage the tube due to heating, which is best carried out using only the hot air of the flame.

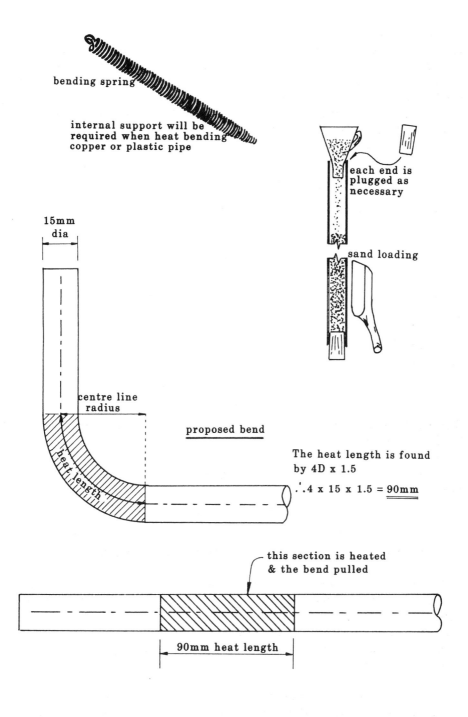

bending spring

internal support will be
required when heat bending
copper or plastic pipe

each end is
plugged as
necessary

sand loading

15mm
dia

centre line
radius

heat length

proposed bend

The heat length is found
by 4D x 1.5

∴.4 x 15 x 1.5 = 90mm

this section is heated
& the bend pulled

90mm heat length

Pipe Bending Using Heat

Soldering

This is a process in which alloys are used to join metals below their melting temperature – unlike welding, in which the joint is made by fusing with the parent metal. In order to solder successfully proceed as follows:

(1) Thoroughly clean the joint;
(2) Apply flux (it should be noted that some soldering processes using an oxyacetylene flame do not require a flux);
(3) Apply heat (either directly or via a soldering iron);
(4) Feed solder to the joint;
(5) Allow the joint to cool, without movement;
(6) Remove excess flux. Failure to do so could result in corrosion problems.

Many soldered joints, be they soft or hard, are formed by allowing the solder to flow between two close surfaces; these joints are called capillary or brazed joints. A brazed joint is one which remains strong at high temperatures (i.e. hard soldered). Capillary joints can be made at any angle because the solder does not fill the joint like filling a cup; it is drawn between the two close-fitting surfaces.

The correct flux must always be used, as specified by the manufacturer. It is applied to prevent the oxygen in the atmosphere coming into contact with the cleaned metal and thus prevents oxidation.

Soft soldering The term soft soldering refers to any soldering process in which the solder used makes a joint which will not withstand too much stress and generally soft-soldered joints are made with a flame temperature no higher than about 450°C.

Soft solders include lead/tin and lead-free solders which are composed mainly of tin and small amounts of copper, silver or antimony to give strength.

Hard soldering There are several types of hard solder including silver, or silver alloys with varying percentages of copper, or the cheaper-to-purchase copper/phosphorus alloys (cupro-tected). Hard-soldered capillary joints are made in much the same way as soft-soldered joints, the main difference being the temperatures required which range from 600°C to 850°C; because of these temperatures a very secure job is made.

Bronze welding is also a form of hard-soldered joint (see page 50).

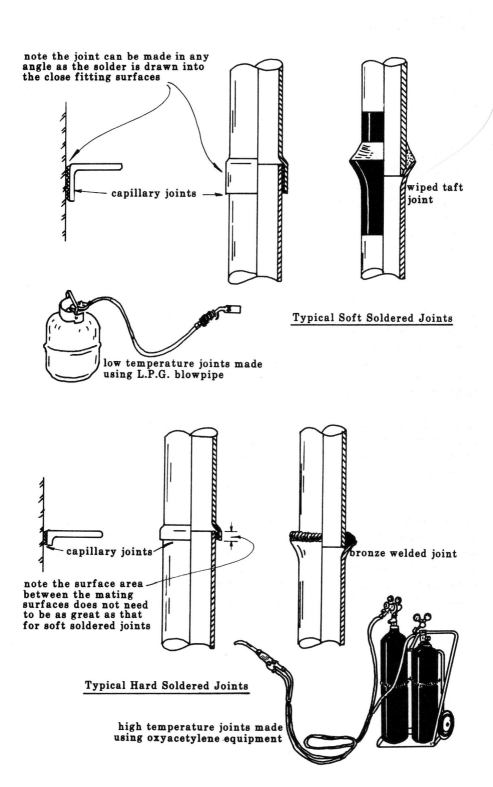

note the joint can be made in any angle as the solder is drawn into the close fitting surfaces

capillary joints

wiped taft joint

Typical Soft Soldered Joints

low temperature joints made using L.P.G. blowpipe

capillary joints

bronze welded joint

note the surface area between the mating surfaces does not need to be as great as that for soft soldered joints

Typical Hard Soldered Joints

high temperature joints made using oxyacetylene equipment

Soldering

Welding Equipment and Safety

Relevant British Standards
BS 6503 and BS 6158

The gases most commonly used in welding processes include oxygen, which is supplied in a black cylinder with a right-handed thread for the connection of the regulator; and acetylene, which is supplied in a maroon-coloured cylinder with a left-handed thread; the oxygen hose is blue or black and the acetylene hose red. To the other end of the hoses is fitted the required blowpipe with the correct welding nozzle.

When carrying out any welding processes you should take several precautions to ensure your own safety and that of those around you; the following checklist should be observed:

☐ Always wear protective clothing, especially the correct eye goggles or shields.
☐ At all times ensure good ventilation when welding.
☐ Erect any necessary signs or shields to give protection to people and to warn them of the welding process taking place.
☐ Always have fire-fighting apparatus to hand.
☐ Always repair or replace perished or leaking hoses with the correct fittings; do not use odd bits of tubing to join the hoses. On no account should piping or fittings made of copper be used with acetylene, as an explosive compound would be produced. Acetylene should never be allowed to come into contact with an alloy containing more than 70% copper.
☐ Store the gas cylinders in a fireproof room. If it can be avoided, oxygen should not be stored with combustible gases such as acetylene; full and empty cylinders should be kept apart.
☐ Acetylene cylinders should always be stored and used upright to prevent any leakage of the acetone.
☐ Oxygen cylinders must not be allowed to fall because, should the valve be broken off, the high pressure in the cylinder could cause the cylinder to shoot off like a torpedo, causing extensive damage; for this reason oxygen cylinders should always be secured. Oil or grease will ignite violently in the presence of oxygen; therefore cylinders should be kept clear of such materials.
☐ Allow an adequate flow of fuel gas to discharge from the blowpipe before lighting up.
☐ In the event of a serious flashback or backfire plunge the blowpipe in a pail of water to cool it, leaving the oxygen running to prevent water entering the blowpipe.
☐ Hose check valves should be fitted on the blowpipe to prevent any flashback into the hoses; also flashback arrestors should be fitted to the regulators to prevent a flash-back occurring within the cylinder itself.

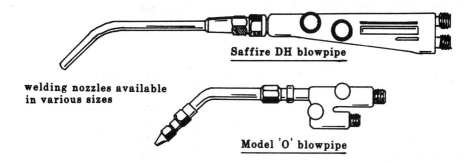

Saffire DH blowpipe

welding nozzles available
in various sizes

Model 'O' blowpipe

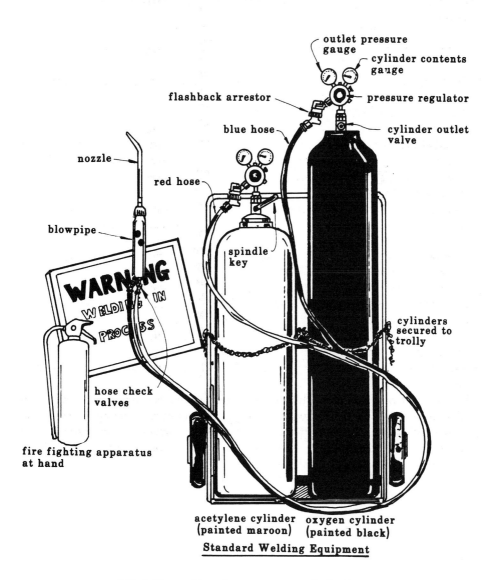

outlet pressure
gauge

cylinder contents
gauge

flashback arrestor

pressure regulator

blue hose

cylinder outlet
valve

nozzle

red hose

blowpipe

WARNING
WELDING IN
PROCESS

spindle
key

cylinders
secured to
trolly

hose check
valves

fire fighting apparatus
at hand

acetylene cylinder
(painted maroon)

oxygen cylinder
(painted black)

Standard Welding Equipment

Welding Equipment and Safety

Welding Processes

Relevant British Standard
BS 499

Welding can be defined as the coalescence, or joining, of metals using heat with or without the application of pressure, and with or without the use of a filler rod. The heat required to weld successfully may be supplied by either electricity or a gas flame. Before modern methods of welding, metals were joined by heating them in a forge to welding temperature and hammering or pressing the two metals together.

With oxyacetylene welding in which oxygen and acetylene gases are used in approximately equal volumes, a flame temperature of about 3200°C is produced at the tip of the inner cone. The flame can be adjusted with an excess amount of oxygen or acetylene which alters its flame characteristics (see figure). Most welding processes, including lead welding, require the use of a neutral flame, bronze welding (see page 50) requiring a slightly oxidising flame.

Successful welding may be carried out by employing either of two basic welding techniques: the leftward or rightward methods. With the leftward welding technique the filler rod precedes the blowpipe and the weld progresses from right to left with the blowpipe nozzle pointing in the direction of the unwelded surfaces. Conversely, with the rightward welding technique the welding progresses in the opposite direction, i.e. from left to right. In order to achieve this and weld successfully the blowpipe nozzle is directed into the completed weld at a much lower angle; also, with this technique the blowpipe flame precedes the filler rod along the weld.

A 'true' welded joint uses a filler rod of the same metal as that of the metals being welded, and is referred to as an autogenous weld. Welding processes such as bronze welding are not in fact welded joints but are a process of hard soldering.

Welding is carried out by first ensuring that the edges of the metal are free from oxides, dirt or grease, etc., and the metal butted together. The correct blowpipe is chosen, with the correct nozzle, and the flame adjusted to a pressure of approximately 0.14–0.21 bar (2–4 lb/sq in) depending on the metal thickness. The metal is heated to its melting temperature and the filler rod applied as necessary. Thick steel will require higher gas pressures than those suggested.

When producing an autogenous weld on such metals as lead or steel, using a neutral flame, no flux is required because the oxyacetylene flame prevents the gases in the air coming into contact with the surface metal, and oxidation will therefore not result. This is not to be taken as the rule, as some welding processes, e.g. those employed when welding aluminium, will require a flux.

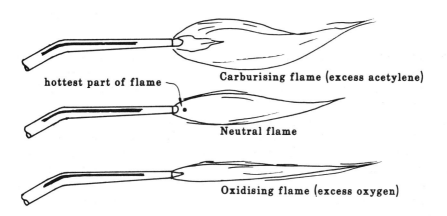

hottest part of flame

Carburising flame (excess acetylene)

Neutral flame

Oxidising flame (excess oxygen)

The Oxyacetylene Flame

filler rod

nozzle

60-70°

30-40°

direction of weld

blowpipe is generally directed
from side to side

Leftward welding technique

filler rod

30-40°

40-50°

direction of weld

filler rod follows the blowpipe
in a circular motion

Rightward welding technique

Welding Processes

Bronze Welding

Relevant British Standard
BS 1724

Bronze welding is used to join various metals; the resulting join is referred to as a welded joint but in fact no melting of the parent metal takes place; therefore it is a type of hard-soldered joint. To make a bronze welded joint a space is required between the two surfaces to be joined; this can be achieved by grinding thick material back to form a bevelled edge or in the case of pipe a space could be left or a bell joint formed. The joint should be thoroughly cleaned to remove any oxide coating, then heat applied to the joint with a slightly oxidising flame, which is worked from side to side or in small circles over the surface.

A slightly oxidising flame is chosen because the excess oxygen in the flame is required to mix with the zinc in the filler rod, which boils and changes to a vapour below the melting temperature of copper. When the oxygen mixes with the zinc a zinc oxide is formed which melts at a much higher melting temperature (pure zinc oxide melts at 1975°C); failure to observe this will result in the zinc changing to a vapour and bubbling up through the weld, leaving a series of blowholes.

The fluxed filler rod is introduced to the joint and the flame is lifted to stroke the rod which, upon melting, should adhere to the metal being joined. The process is continued slowly, progressing along the joint, adopting the leftward welding technique (see page 48) forming a series of characteristic weld ripples. The filler rods mainly used for bronze welding consist of approximately 60% copper and 40% zinc with a small amount of tin and silicon to act as deoxidisers and assist in its flowing characteristics.

One of the advantages of bronze welding is the temperature at which the filler rod melts (around 850–950°C); this minimises the distortion which would otherwise take place should the joint be fusion-welded, requiring a temperature of at least the melting point of the parent metal. Also, bronze welding can be used to join dissimilar metals, such as copper and iron.

A special flux is used to bronze weld: borax and silicon; this is either made into a paste by mixing with water and applied directly to the joint, or a heated filler rod is dipped into the powdered flux which in turn sticks to the rod. Special flux-impregnated filler rods can be purchased at a little extra cost. As with all joints requiring a flux, ensure that any excess flux is removed on completion of the weld. It is sometimes possible to loosen the flux residue by quenching in water immediately after bronze welding the joint.

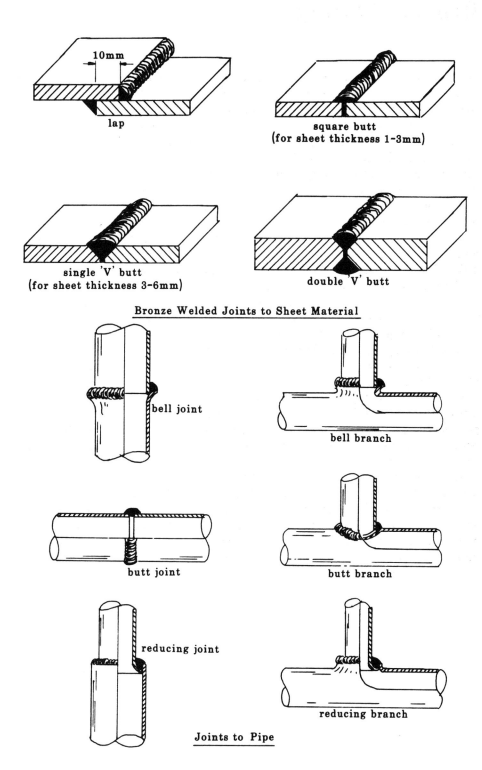

10mm

lap

square butt
(for sheet thickness 1-3mm)

single 'V' butt
(for sheet thickness 3-6mm)

double 'V' butt

Bronze Welded Joints to Sheet Material

bell joint

bell branch

butt joint

butt branch

reducing joint

reducing branch

Joints to Pipe

Bronze Welding

Lead Welding (Lead Burning)

Lead can easily be welded together with either oxyacetylene welding equipment, or propane gas, although propane gas welding is limited to simple butt and lap joints. Filler rods are generally made from strips of sheet lead 6–8 mm wide, cut from waste. One of the prerequisites of successful welding is cleanliness of the filler rod and the surfaces to be welded. The correct nozzle must be chosen (model '0' blowpipe nos 2–3) and adjusted to a neutral flame (see page 48). The pressure at both oxygen and acetylene regulators should be adjusted to 14 kN or 0.14 bar (2 lb/sq in.). A finished weld should look uniform in size and shape and it is essential that there is sufficient penetration; equally there should be no undercutting or overlapping to the edge of the weld.

Butt welds When a butt weld is to be carried out the meeting surfaces should first be shaved clean, approximately 6 mm wide on each sheet. Then a tack weld is applied at several intervals along the joint; this prevents the joint opening because of expansion. The welding nozzle is held close to the joint at an angle of about 60° and a molten pool is allowed to become established; the filler rod is introduced close to the nozzle and the blowpipe is slightly raised to melt off a piece of the filler rod which drops and fuses with the molten pool. The flame is returned to the pool in a stroking action (see figure). The blowpipe is then moved forwards to a new position where the process is repeated.

Lap welds The lap joint is formed by carring out a welding process similar to that for the butt joint, except that one sheet (the overcloak) is lapped on top of the sheet to be joined by 25 mm. To prepare this joint the overcloak is cleaned on both sides, placed on the cleaned undercloak and tacked in position. The first weld is made, joining the two sheets; this causes a certain amount of undercutting to the overcloak, so that a second weld is required as reinforcement.

Vertical welds With vertical welding no filler rod is used and the joint is prepared as for lap welding. The nozzle is held close to the overcloak and as the lead begins to melt the nozzle is circled round like a No 6 finishing off at the undercloak; a molten lead bead will follow the flame and when it becomes fused with the undercloak the flame is removed.

Inclined welds Two different techniques can be adopted to create lead welding up an incline; one produces a true incline joint in which the seams to be welded are either butted or lapped together and welding of the sheet is carried out by depositing the lead up the incline in the form of overlapping semicircular beads. The second technique really produces a vertical welded joint in which the two adjoining sheets overlap each other at an angle; lead welding vertically in this fashion tends to be easier and stronger than a true vertical lead weld.

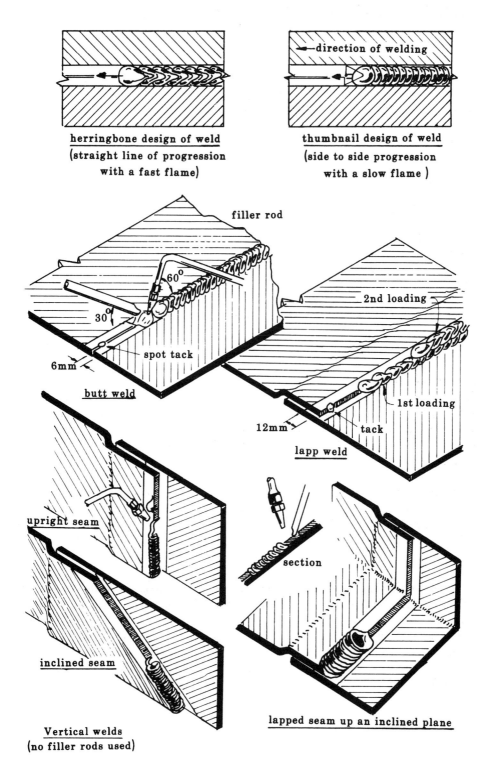

herringbone design of weld
(straight line of progression
with a fast flame)

direction of welding

thumbnail design of weld
(side to side progression
with a slow flame)

filler rod

60°

30°

6mm

spot tack

butt weld

2nd loading

1st loading

12mm

tack

lapp weld

upright seam

inclined seam

Vertical welds
(no filler rods used)

section

lapped seam up an inclined plane

Lead Welding

Cocks, Taps and Valves 1

Relevant British Standards
BS 1010, BS 1552 and BS 5433

The definition of a cock, tap or valve can give rise to many an argument; in general, I suggest, the differences are just matters of local terminology: e.g. bib-tap and bib-cock or stop-tap and stop-cock. The purpose of a valve is to adjust and regulate the velocity and flow through a pipeline, either in line, or at the point of termination.

Valves fitted in-line

Screwdown valves Several designs of in-line screwdown valve will be found to include the stop-cock and disc globe cock. Screwdown valves are characterised by a plate or disc, shutting slowly, at right angles to the valve seating. Stop-cocks are used on high pressure pipework and have a rubber washer fitted to give a sound seal. Note the direction in which the valve must be fitted so that it will lift the washer off the seating should it become detached from the jumper; failure to do so may result in a no-flow situation.

The disc globe cock is similar in construction to the stop-cock although more robust in its construction and used in high pressure- or temperature pipework.

Gatevalves This valve is sometimes referred to as a fullway gatevalve because when it is fully open there is no restriction of flow through the valve, unlike the screwdown valve where the liquid passes up and through a seating aperture. For this reason gatevalves are recommended for use where there is low pressure in the pipeline, such as when fed from a storage cistern. It is recommended that these valves be fitted in vertical pipework to prevent sludge settling in the base of the seating and thus stopping the gate from fully closing.

Quarter-turn valves Two main types of quarter-turn valve will be found: the plug-cock and the ballvalve. These valves work by the quarter turn operation of a square head located at the top of the valve, which aligns a hole in the valve with that of the pipe. A slot will be noticed in this head and when it is in line with the direction of the pipe it indicates that the valve is open. The plug-cock has traditionally been used for gas pipelines and is widely known as the 'gas-cock'. The plug-cock has a tapered shaped plug whereas the ballvalve consists of a circular ball through which the liquid or gas can flow, when the valve is open. The term *ballvalve* is rarely used (thus it is seldom confused with float-operated valves), and trade names such as Ballafix or Minuet are used instead. These valves are very common today and are used for a number of operations such as the service valve to a cistern or washing machine, some being fitted with a turning handle whilst others need the use of a screwdriver. Large ballvalves are sometimes installed on cold feed and distribution pipework, replacing the more traditional gatevalve, being more effective in their operation.

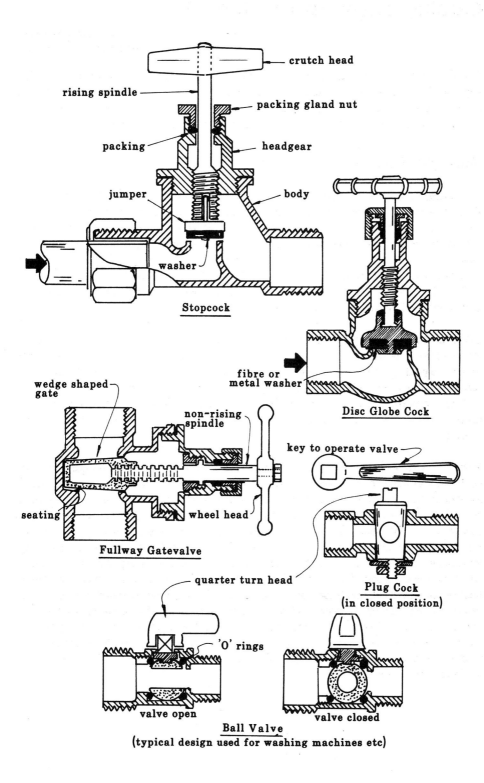

Cocks, Taps and Valves 1

Cocks, Taps and Valves 2

Relevant British Standard
BS 1010

Terminal valves

A valve which is fitted at the point of use is sometimes called a terminal fitting, a reference to the point where it 'terminates'. Several terminal valves are available, including draw-off taps, drain-off cocks and float-operated valves.

Draw-off taps generally work in a similar fashion to screwdown valves (previously mentioned). Many modern valves make use of ceramic discs; these do not work as screwdown valves but allow two polished ceramic discs to turn and align up two port holes through which the water can pass. These are very popular at present, allowing for a quarter turn design of tap to be used, but caution must be given to water hammer problems which may arise.

The **bib-tap** is characterised by a horizontal male iron thread which is screwed directly into a fitting. A typical location example would be above a butler sink or fitted outside for use with a hose pipe connection.

The **piller-tap** differs from the bib-tap in that it has a long vertical male iron thread which passes through the appliance and is held in position with a back-nut, the supply being fed to this point below the appliance.

The **globe-tap** is now obsolete, but will occasionally be found. It had a female thread as its point of entry and was used for old baths with a side entry. Installation of this valve is no longer permitted owing to back-siphonage problems.

Mixer taps are designed to allow the flow of water from hot and cold supplies to be delivered via one outlet spout. Depending on the design of mixer the water either mixes in the body of the tap or as it comes out of the spout (see figure). It is a requirement that waters under different pressures be mixed as they leave the spout, otherwise it would not be possible to get water from the supply under the least pressure. A monoblock is simply a design of mixer where only one hole is required in the appliance, designed for cosmetic reasons.

The **supatap**, available in both 'bib' or 'pillar' design, offers a facility whereby it is possible to re-washer the tap without having to turn off the water supply. This is achieved by an automatic closing valve which drops to stop the flow of water should the nozzle be removed.

A **drain-off cock** is a valve located at the lowest point of any system and has a serrated hose connection outlet. Drain taps fitted to supply pipes must conform to BS 2879 and be of the screwdown pattern.

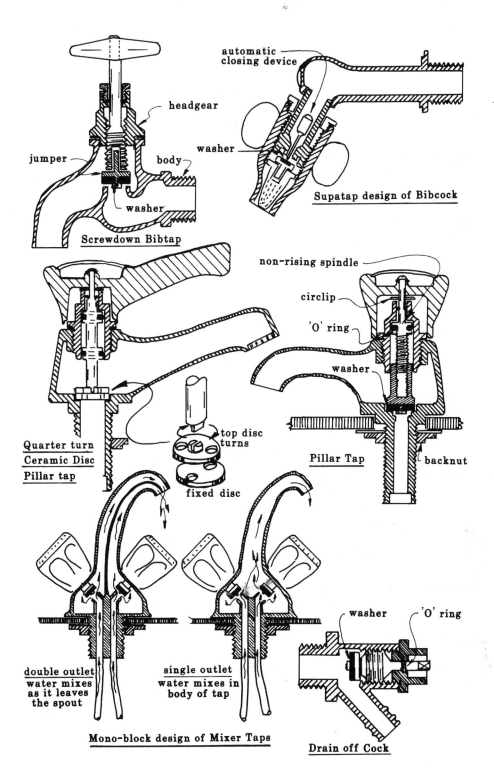

headgear

jumper

body

washer

Screwdown Bibtap

automatic closing device

washer

Supatap design of Bibcock

Quarter turn Ceramic Disc Pillar tap

top disc turns

fixed disc

non-rising spindle

circlip

'O' ring

washer

Pillar Tap

backnut

double outlet water mixes as it leaves the spout

single outlet water mixes in body of tap

Mono-block design of Mixer Taps

washer

'O' ring

Drain off Cock

Cocks, Taps and Valves 2

Cocks, Taps and Valves 3

Relevant British Standard
BS 1212

Float-operated valves

These are commonly called 'ballvalves', owing to the use of a ball float, although the name is somewhat inappropriate with some modern designs. These valves are fitted into cisterns in order to automatically regulate the water supply. The valve closes as the water reaches a predetermined level. It works on the principle of leverage; as the water rises it causes the ball to become buoyant, exerting an upward thrust which is transferred, via the lever and its fulcrum, to the plunger or piston; this gradually closes the washer onto the seating. The valve is available in several designs: the diaphragm; Portsmouth or Croydon.

The **diaphragm ballvalve** consists of a large diaphragm washer which is forced onto a seating by a small plunger. This design conforms to BS 1212, parts 2 or 3, and has been the only design permitted in domestic premises since the 1989 local water byelaw; this is because it is able to maintain a type 'B' air gap and therefore prevent back-siphonage.

The **Portsmouth ballvalve** is commonly seen, although it is now being phased out in favour of the diaphragm valve. The Portsmouth valve is more sluggish in use than the latter and no longer conforms to current water regulations.

The **Croydon ballvalve** is now obsolete; it worked in a similar fashion to the Portsmouth, although the piston moved up and down instead of from side to side.

Some designs, referred to as equilibrium valves, allow water to pass through the washer to its other side; thus the water pressure acting on the valve is in a state of equilibrium (pushing on both sides of the washer). The float of this type of valve only has to lift the weight of the arm, whereas in the case of simple lever types the effort provided by the float not only has to overcome the weight of the arm but also the pressure of the incoming water.

Equilibrium valves prove useful in areas where the water supply is very high or where persistent water hammer may be encountered.

It is possible to purchase ceramic disc-float-operated valves as with draw-off taps, previously mentioned.

It is a byelaw requirement that whenever a float-operated valve is fitted a means of servicing the valve must be provided, which is achieved by the installation of a service valve fitted as close as practically possible to the cistern in which the float-operated valve is fitted.

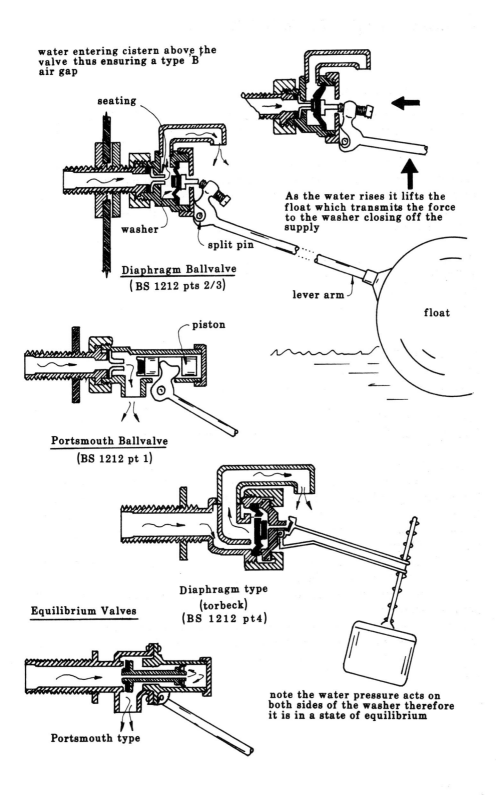

water entering cistern above the valve thus ensuring a type 'B' air gap

seating

washer

split pin

Diaphragm Ballvalve
(BS 1212 pts 2/3)

As the water rises it lifts the float which transmits the force to the washer closing off the supply

lever arm

float

piston

Portsmouth Ballvalve
(BS 1212 pt 1)

Equilibrium Valves

Diaphragm type
(torbeck)
(BS 1212 pt4)

Portsmouth type

note the water pressure acts on both sides of the washer therefore it is in a state of equilibrium

Cocks, Taps and Valves 3

Part 2
Hot and Cold Water Supplies

Classification of Water

Water may be produced as a result of burning hydrogen. For example, every day we burn natural gas, which is one form of hydrocarbon (CH_4); this gives off (as its by-products) hydrogen oxide (H_2O, i.e. water) and carbon dioxide (CO_2).

Pure water is a transparent, tasteless liquid and will be found in three physical states: solid (ice); liquid; or gas (steam or vapour). At atmospheric pressure water will be found as a liquid in the temperature range 0–100°C. At 0°C water changes to ice with an immediate expansion in volume of 10%. At 100°C it changes to steam, its volume expanding some 1600 times.

To convert water back to its element gases an electric current needs to pass through the liquid.

Water is rarely 'pure' in its true sense because it is contaminated with gases or chemicals which it absorbed as it fell as rain. Generally speaking water is classified as either hard or soft, this classification referring to the parts of calcium it contains per volume.

Soft water

This is water which is free from dissolved salts such as calcium carbonates and sulphates. Naturally occurring soft water is slightly acidic due to absorbed gases such as CO_2. Soft water tends to be more pleasant to the skin and to wash in, but has the major disadvantage of being aggressive to pipework, in particular lead.

Hard water

This is water which has fallen on and filtered through calcium carbonates or sulphates (chalk or limestone). It dissolves the calcium, carrying it in suspension. The water may be either permanently or temporarily hard.

Permanent hardness This is the result of water absorbing calcium or magnesium sulphates, which dissolve due to the natural solvency of water.

Temporary hardness This is the result of the water absorbing calcium carbonates. As rain falls it absorbs CO_2; water containing CO_2 dissolves the calcium carbonates in the ground and converts them to soluble bicarbonates. These bicarbonates are then carried in suspension. When the water is heated in excess of 65°C the water molecules jump about so much that the CO_2 gas escapes from the water, back into the surrounding air. Once the CO_2 has gone the bicarbonate reverts back to its original form as a calcium carbonate and is precipitated in the form of scale, accumulating inside the system.

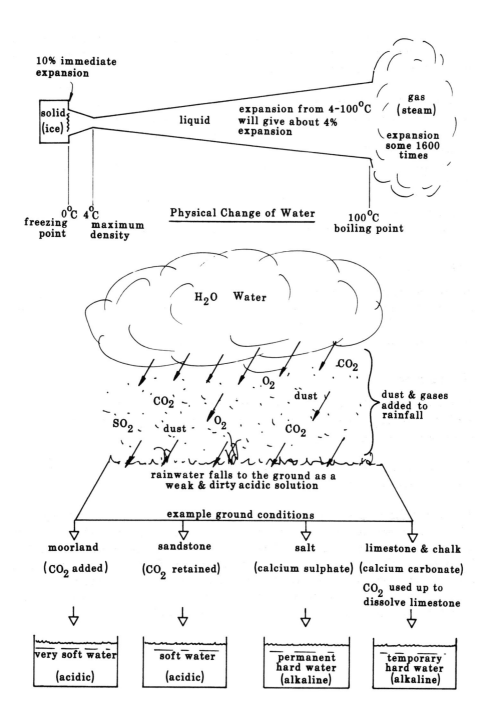

Classification of Water

Cold Water to the Consumer

Relevant British Standard
BS 6700

Throughout the United Kingdom potable water (i.e. water fit for human consumption) is provided by the local water authorities to individual premises and various industries. When a supply of water is required the water authority will supply water to a point just outside the property boundary line where nowadays a meter is usually installed to calculate the water consumed. From this point the supply pipe is run into the premises, with precautions being taken to protect the pipe from movement, frost damage and aggressive soil.

Any pipe passing through or under a building must be ducted; this allows for its removal should the need arise. A consumer's stop tap is fitted where the supply pipe enters the building and should be fixed as low as possible with a drain-off cock immediately above it. Note that older properties do not have meters fitted. The pipe is run from this point to feed the various systems of cold water supply.

Local water byelaw requirements

There are some 100 local water byelaws which need to be observed when connecting to the local water supply main; these are primarily designed to prevent the wastage and contamination of water.

Wastage of water This could be the result of undue consumption, misuse or simply a faulty component, such as a leaking valve. To prevent this, water meters are being installed to register the amount of water used in serving the premises, for which the owner/occupier will eventually be invoiced. As a means of combating this problem the installer must provide the dwelling with suitable overflow/warning pipes from cisterns to let the user know of a fault, and a means of isolation must be provided, to allow its speedy shutdown. Prevention from damage must also be maintained by allowing for thermal movement, frost damage, etc.

Water contamination There are several means by which water may be inadvertently contaminated; for example, it is possible that water may be sucked up from sanitary appliances (back-siphonage), such as baths and sinks; this is caused by peak flow demands creating a negative pressure on the supply pipe. If the supply main were to have a pressure less than the hot supply at mixer valves, again a back-flow would occur. The installation must be designed to prevent this back-flow by maintaining a suitable air gap between the appliance and the outlet spout; alternatively one must provide some other suitable means of back-flow prevention, such as a double check-valve assembly (i.e. two non-return valves fitted in the line).

Contamination can also result from the use of unsuitable materials; lead, for instance, is dissolved by water and as a result is prohibited from use nowadays; even the solder used to join copper pipes, when used for hot or cold supplies, must be lead free.

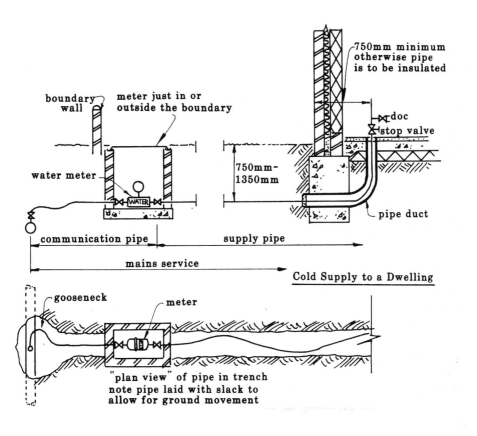

boundary wall

meter just in or outside the boundary

750mm minimum otherwise pipe is to be insulated

doc

stop valve

water meter

WATER

750mm-1350mm

pipe duct

communication pipe

supply pipe

mains service

Cold Supply to a Dwelling

gooseneck

meter

"plan view" of pipe in trench note pipe laid with slack to allow for ground movement

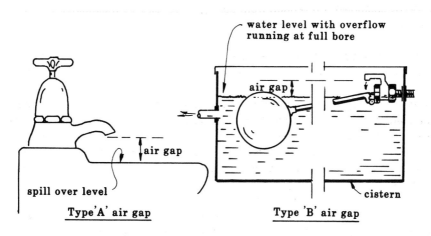

water level with overflow running at full bore

air gap

air gap

spill over level

Type 'A' air gap

cistern

Type 'B' air gap

Backflow Prevention

Cold Water to the Consumer

Cold Water Systems

Relevant British Standard
BS 6700

Two distinct systems of cold water supply are in use: the direct and indirect systems, although modified systems are to be found, in which several appliances are on the mains supply and several fed from a cistern. It is essential that the plumber obtains advice and gives written notice of the design of a new system to the local water authority before commencing work; failure to do this may mean a contravention of the local water byelaws.

Direct system

In this system all the cold water in the house is fed 'directly' from the supply main. The water pressure is usually high at all outlets, so this system can have the disadvantage of being more prone to water hammer. In some areas the supply pressure is reduced at peak times; this can cause a negative pressure in the pipeline and loss of supply. Also, precautions need to be taken to prevent back-siphonage of fouled water from appliances into the supply pipe.

The direct system is cheaper to install than the indirect, and allows for a house design which does not require a roof space to accommodate the cistern, but peak flow times must be considered and above all adequate pipe sizing to prevent a lack of suitable flow, should several appliances be used at once.

Indirect system

In this system only one draw-off point (i.e. the kitchen sink) is fed from the mains supply pipe; all other outlets are supplied via a cold storage cistern, usually located in the roof space. Water pressure is usually much lower than with the direct system, but it will be maintained, even at peak times or during complete shutdown of the supply. Today in modern housing, with suitable precautions to storage cistern sizing and the prevention of water contamination, stored water is regarded as potable (fit for human consumption); therefore drinking supplies are not only regarded as on the 'mains' supply pipe; to this end the same precautions must be maintained to prevent back-siphonage.

Cold supply to the domestic hot water (dhw)

Vented systems These require a supply via a cold feed cistern. The cold feed pipe is run separate from any cold distribution pipework, to prevent hot water being drawn off when the cold supply is opened.

Unvented systems These systems are fed directly from the mains supply pipe. The biggest consideration is that the supply main is large enough in diameter to provide a good flow rate, should several appliances be opened at once. To prevent the hot water back-flowing down the feed pipe a check-valve must be incorporated (see Unvented Domestic Hot Water Supply, page 82).

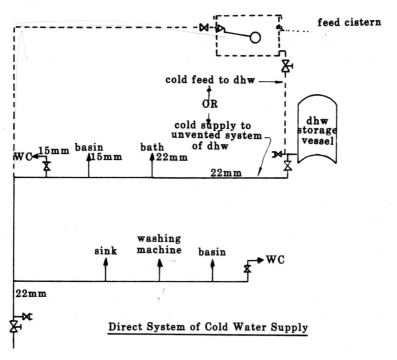

feed cistern

cold feed to dhw

OR

cold supply to
unvented system
of dhw

dhw
storage
vessel

WC 15mm basin bath
15mm 22mm

22mm

washing
sink machine basin
WC

22mm

Direct System of Cold Water Supply

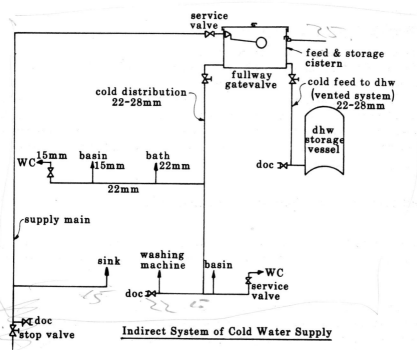

service
valve

feed & storage
cistern

fullway
gatevalve

cold distribution
22-28mm

cold feed to dhw
(vented system)
22-28mm

dhw
storage
vessel

WC 15mm basin bath
15mm 22mm

doc

22mm

supply main

washing
sink machine basin
doc WC
service
valve

doc
stop valve

Indirect System of Cold Water Supply

Cold Water Systems

Cold Water Storage

Relevant British Standard
BS 6700

When cold water is required to be stored for use as a supply for an indirect system of cold water, or for the feed to a system of dhw, the water is stored in a cistern, usually located in the roof space.

The storage cistern should have a minimum capacity of 100 litres. If the cistern is also to act as a feed cistern for the hot water supply, (being a combined storage and feed cistern) it should have a minimum capacity of 230 litres.

Cold distribution pipes from storage cisterns should be connected so that the lowest point of the water outlet is a minimum of 30 mm above the base of the cistern; this is to prevent sediment passing into the pipework. Connections of feed pipes to hot water apparatus from cisterns should be at least 25 mm above cold distribution pipes, if applicable.

This should minimise the risk of scalding should the cistern run dry. The supply ballvalve is fitted as high as possible and must comply with BS 1212, parts 2 or 3, thus maintaining a type 'B' air gap and preventing back-siphonage. Overflow pipes should have a minimum internal diameter of 19 mm and in all cases be greater in size than the inlet pipe.

To prevent the ingress of insects a tight-fitting lid must be provided, with a screened air inlet; where a vent pipe passes through the lid the pipe must be sleeved; overflow/ warning pipes must also incorporate a filter or screen. Finally the whole installation (cistern and pipes) must be insulated to prevent freezing.

Coupling of storage cisterns

Sometimes it is desirable to have two or more cisterns coupled together instead of one large cistern. This is beneficial because one of them can always be isolated and drained down, if required. Equally, on a smaller scale, in a house, lack of space sometimes limits the size of a storage cistern; once again two smaller cisterns can be joined together to give the required capacity. Different methods are used for the above examples. If means are required to isolate one cistern an isolating valve is fitted at each point in or out of the cistern, which can be shut off. For the purposes of a domestic house it would be uncommon for one cistern to be isolated, so the mains supply is usually taken into one cistern and the cold distribution or feed pipe is taken out of the other. By designing it in this way the water in the second cistern would not become stagnant.

The washout pipe shown in the figure is only used on large cisterns (those holding over 2300 litres), for the purpose of draining down and cleaning out any sludge deposits, etc.

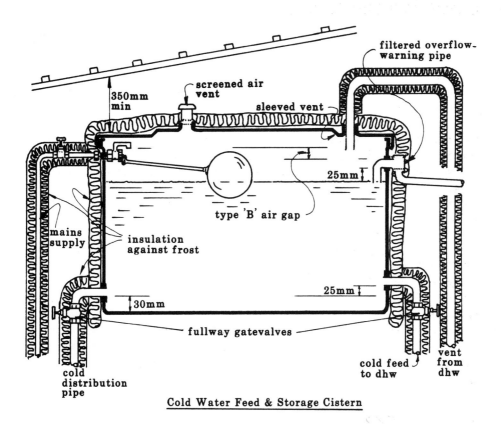

filtered overflow-
warning pipe

screened air
vent

350mm
min

sleeved vent

type 'B' air gap

25mm

mains
supply

insulation
against frost

25mm

30mm

fullway gatevalves

cold feed
to dhw

vent
from
dhw

cold
distribution
pipe

Cold Water Feed & Storage Cistern

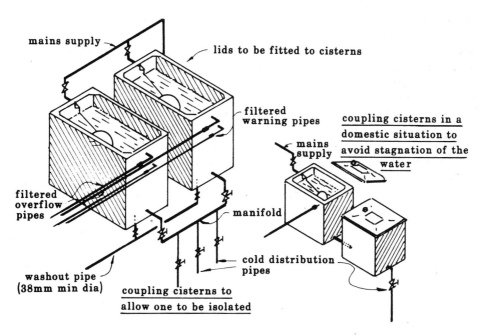

mains supply

lids to be fitted to cisterns

filtered
warning pipes

coupling cisterns in a
domestic situation to
avoid stagnation of the
water

mains
supply

filtered
overflow
pipes

manifold

cold distribution
pipes

washout pipe
(38mm min dia)

coupling cisterns to
allow one to be isolated

Cold Water Storage

Water Treatment

Before water is supplied to the consumer it is treated/purified by the water authority; when it arrives no more treatment, generally speaking, is required. In hard water areas, where there are varying amounts of calcium salts in the water, it is sometimes desirable to treat the water in order to prevent excessive scale problems or provide a better liquid with which to wash (e.g. a laundry). The installation of a water softener or a water conditioner fulfils this purpose.

Water softeners

These provide a water softening process whereby hard water is allowed to pass through a pressure vessel containing zeolites, or a resin which absorbs the calcium in the water. After a period of time the zeolites become exhausted owing to the fact that they become clogged with calcium; they thus need to be regenerated with common salt. This is done automatically, by a system of backwashing which is timed to operate at around 3 AM via a timeclock or flow metering system; thus there is no inconvenience to the householder. Prior to the installation of the softener inlet connection a branch pipe should be taken from the mains to provide a hard water drinking supply.

The installation of a water softener is quite straightforward, the connections being made in the way shown in the figure.

Water conditioners

Water conditioners do not soften water; they just stabilise the calcium salts which are held in suspension. There are two basic types: those that use chemicals and those that pass water through an electronic or magnetic field. The calcium salts, if viewed through a microscope, appear star-shaped, and it is in this form that they can bind together.

Chemical water conditioners use polyphosphonate crystals; these dissolve into the cold water and bind onto and in the star-shaped salts, making them circular. These polyphosphonate crystals are placed in the storage cistern or into specially designed containers fitted into the pipeline. Periodically the crystals must be replaced.

Electronic and magnetic water conditioners are devices fitted in the pipeline which pass a low current of a few milliamperes of electricity across the flow of water; this tends to alter the structure of the hard salts, making them round- or square-shaped. As can be seen, both methods make the hard salts round in shape. In this form they do not stick together but should pass through the system. Electronic and magnetic water conditioners should be installed as close as possible to the incoming main supply. Some types of electronic conditioners are plugged into the mains electric supply, whereas others rely on the current produced by electrolysis.

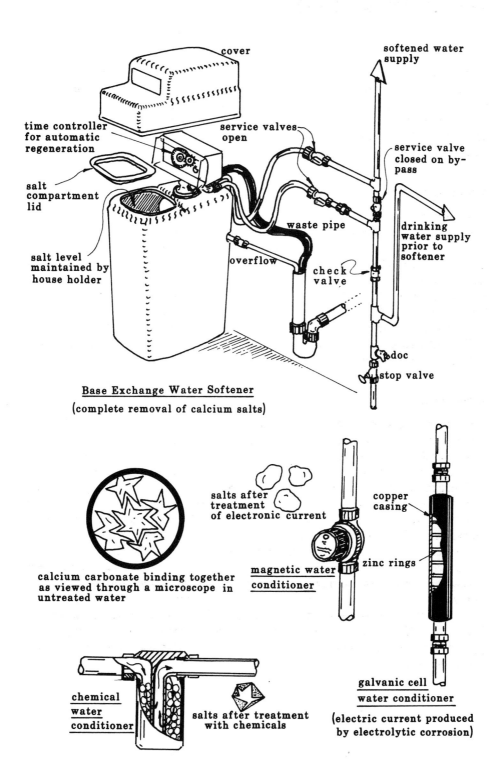

cover

softened water
supply

time controller
for automatic
regeneration

service valves
open

service valve
closed on by-
pass

salt
compartment
lid

drinking
water supply
prior to
softener

waste pipe

salt level
maintained by
house holder

overflow

check
valve

adoc

stop valve

Base Exchange Water Softener
(complete removal of calcium salts)

salts after
treatment
of electronic current

copper
casing

zinc rings

calcium carbonate binding together
as viewed through a microscope in
untreated water

**magnetic water
conditioner**

**chemical
water
conditioner**

salts after treatment
with chemicals

**galvanic cell
water conditioner**

(electric current produced
by electrolytic corrosion)

Water Treatment

Boosted Water Supplies

Relevant British Standard
BS 6700

There are two reasons why water may need to be boosted: (1) to give a better flow and pressure at the draw-off point in a domestic situation (see page 94 Connections to Hot and Cold Pipework); or (2) as a method of raising the water supply in high-rise buildings above that to which the mains will supply.

The pumps are usually fitted indirectly to the supply main, if fitted directly a serious drop in the mains pressure may occur when the pumps are running. The indirect system consists of a suitably sized break cistern located at the inlet to the pumping set (see figure). Nowadays 'packaged' pumping sets are installed; these basically consist of dual pumps to overcome the problem of failure of (or the need to renew) one of the pumps. The second pump also assists at times of high demands on the system, cutting in as necessary. To prevent pump seizure and stagnation of water the pumps should be designed to work alternately. Two basic systems will be found: those using pressure-sensing devices or those using float switches.

Pressure-sensing devices These include transducers or pressure switches which sense the drop in pressure in the pipeline. These come fitted to, and form part of, the packaged pumping set. To prevent the continuous cutting in and out of the pumps a delayed action ballvalve is used in the high level cisterns. If draw-off points are required on the riser a pressurised pneumatic storage vessel is sometimes incorporated to prevent the continual cutting in and out of the pumps. Basically this consists of a vessel containing a rubber bag surrounded by a charge of air. When the pumps are running, water can enter and fill the bag, taking up volume and compressing the air. When the pumps are turned off the compressed air forces the water back out into the pipeline, as and when required.

Float switches These are devices which rise and fall with the water level; they are therefore located in cisterns or pipelines to sense a drop or lack of water within the system. If the high level cistern is of a large capacity it may be necessary to have a drinking water header to prevent stagnation, or a separate high level cistern for drinking water purposes.

To prevent the pumps running dry for any reason, a sensing device needs to be incorporated in the pipe feeding the pumps, e.g. an in-line sensor or a float switch fitted in the break cistern.

Water hammer can be a problem when fitting pumps to any system; therefore a hydropneumatic accumulator should be incorporated if necessary; in fact, packaged pumping sets incorporate these as standard. They are basically small pressurised pneumatic vessels, the air taking up the shock wave (for example see page 105).

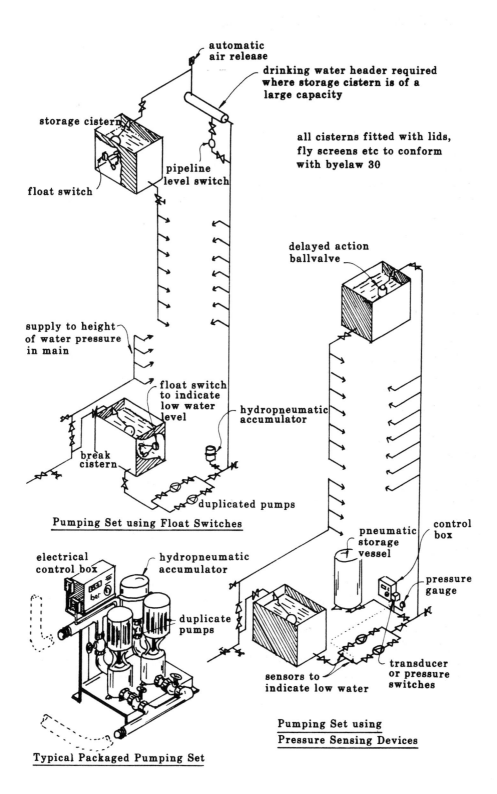

automatic air release

drinking water header required where storage cistern is of a large capacity

storage cistern

all cisterns fitted with lids, fly screens etc to conform with byelaw 30

pipeline level switch

float switch

delayed action ballvalve

supply to height of water pressure in main

float switch to indicate low water level

hydropneumatic accumulator

break cistern

duplicated pumps

Pumping Set using Float Switches

control box

pneumatic storage vessel

electrical control box

hydropneumatic accumulator

pressure gauge

duplicate pumps

sensors to indicate low water

transducer or pressure switches

Pumping Set using Pressure Sensing Devices

Typical Packaged Pumping Set

Boosted Water Supplies

Fire-fighting Systems

Sprinkler systems

On the outbreak of fire a sprinkler system causes an automatic discharge of water to be sprayed, usually from sprinkler heads located near the ceiling. See the figure for a typical arrangement to the pipe layout. Basically there are two designs:

☐ The *wet-pipe system*, in which the sprinkler system is permanently charged with water;
☐ The *dry-pipe system*, in which the sprinkler system is charged with compressed air and is used in unheated buildings where the temperature may fall below 0°C; if they were to be charged with water these pipes would be liable to freeze.

The operating principle of a sprinkler head is that when the temperature around the head rises to a predetermined level a water filled glass bulb breaks, or conversely a solder strut melts, allowing the valve to fall and open.

Hose reel systems

When hose reels are used they should be sited in prominent positions adjacent to exits, so that the hose can be taken to within 6 m of any fire. Two basic designs of hose reels are available: those which automatically turn on as the hose is reeled out and those which need to be turned on at the wall. For the latter type a notice must be provided near the reel indicating the need to turn on the supply. The hose reel must be adequate to supply a minimum of 0.4 litres per second.

The water supplies feeding sprinkler systems and hose reels need to be considered (in this connection see, particularly, BS 5306) but they are usually maintained via a system of boosted water supply.

Wet and dry risers

These systems are for the use of the fire brigade and consist of 100–150 mm diameter pipes run up the building with one or two fire brigade hydrants on each floor. The purpose of this pipe is to save time running canvas hoses up the staircases should the building be on fire. The dry pipe system is used in buildings up to 60 metres in height (20 storeys) and is fitted with an inlet at ground level for the fire brigade to connect to the nearest hydrant. The wet riser is used in taller buildings because the mains pressure would be insufficient to rise to such great heights and is charged with water under pressure by a booster pump capable of delivering 23 litres/s. The water pressure supplied to the hydrants should not exceed 690 kN/m^2, otherwise damage may occur to the hose. Therefore, a pressure relief valve is fitted at the hydrant; this opens if the pressure is too great, the discharging water returning to the break cistern.

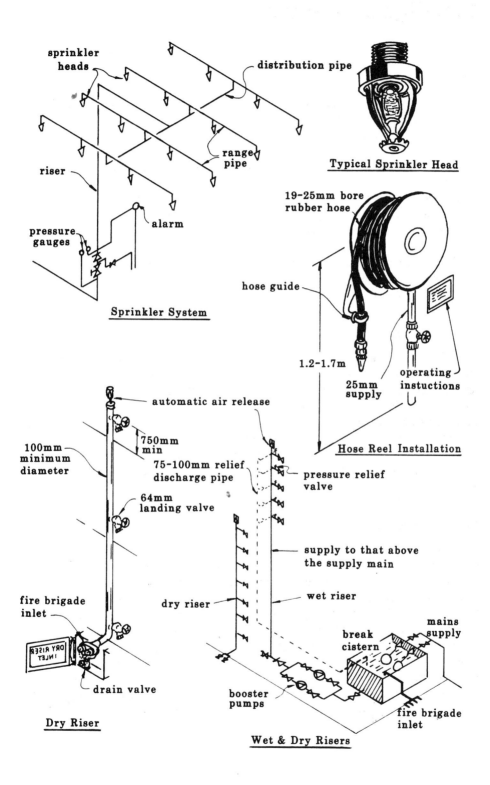

sprinkler heads

distribution pipe

Typical Sprinkler Head

range pipe

riser

alarm

pressure gauges

Sprinkler System

19-25mm bore rubber hose

hose guide

1.2-1.7m

25mm supply

operating instuctions

Hose Reel Installation

automatic air release

750mm min

100mm minimum diameter

75-100mm relief discharge pipe

64mm landing valve

pressure relief valve

supply to that above the supply main

fire brigade inlet

dry riser

wet riser

DRY RISER INLET

drain valve

booster pumps

break cistern

mains supply

fire brigade inlet

Dry Riser

Wet & Dry Risers

Fire-fighting Systems

Hot Water Systems
(Design Considerations)

Relevant British Standard
BS 6700

When a supply of domestic hot water (dhw) is required the designer has to consider many factors to ensure the most suitable system for the building in question.

The chart below gives a brief guide to the system design which could be chosen. Generally dhw can be divided into centralised and localised systems. The terms *instantaneous* and *storage* simply indicate whether the water is heated only as required (instantly) or to a temperature indicated by a thermostat and held in a vessel until required (stored).

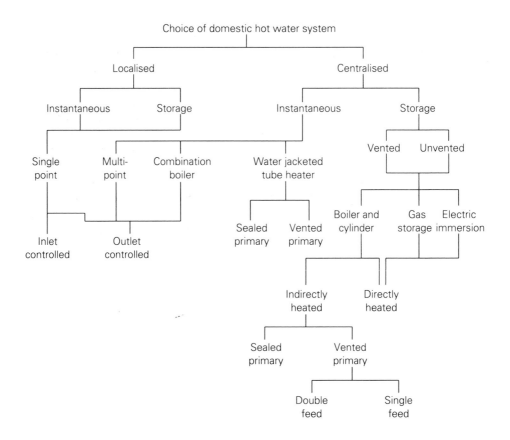

A **centralised system** is one in which the water is heated and possibly stored centrally within the building, supplying a system of pipework to the various draw-off points.

A **localised system** is one in which the water is heated locally to its needs, e.g. a single-point heater located above a sink, and may be chosen where a long distribution pipe would mean an unnecessarily long wait for hot water to be drawn off at the appliance.

In a centralised system the water may be heated in the hot storage vessel itself or it may be heated in a boiler or small gas circulator located at a more convenient position.

Should this be the case the water is fed to and from the boiler by what are known as primary flow and return pipes. The circulation of water can be achieved by *convection currents* being set up in the flow and return pipework or by the use of a circulating pump. Water flows by convection currents as a result of expanding in volume.

As water expands it occupies more space, becoming lighter, volume for volume, than any cooler water. Because hot water rises it is drawn off from the top of the storage vessel to supply the various draw-off points (taps). The cold feed is supplied low down in the vessel, thus preventing unnecessary cooling to the previously heated water. At the highest point in the system a vent pipe is run up to terminate, with an open end just below the feed cistern lid. This pipe is to allow air to escape from the system upon initial filling and allows air in on draining down.

The vent pipe also acts as a fail safe device should the cold feed become blocked, preventing the expanding water passing back up into the cistern; should this occur the water is forced over the vent and discharges into the cistern. The height to which a vent pipe is to rise above the water level in the cold cistern is found by allowing 40 mm for every 1 m head of water in the system, plus an additional 150 mm. For example, if the distance between the base of a boiler and the water level in a cistern is 5 m, the vent pipe should be carried up a minimum distance above the water level of: $5 \times 40 + 150 = 350$ mm.

The temperature at which the water is stored in the cylinder should not exceed 60°C; failure to observe this limit may result in scalding to the user and scale build-up in hard water districts. Some means of controlling the temperature should, therefore, be provided. When a primary circuit is used with an independent boiler, the temperature is generally controlled by (1) a boiler thermostat, (2) a cylinder thermostat operating a motorised valve, or (3) a thermostatic control valve located on the return pipe.

Often the boiler serves the central heating as well as the domestic hot water, and during the winter the boiler thermostat is generally turned up above that used for domestic hot water; therefore the first method ((1) above) often proves unsatisfactory.

Direct Hot Water Supply (Centralised)

Relevant British Standards
BS 5546 and BS 6700

Various mediums and methods of design can be used to heat the water in a centralised system of direct hot water, including the following:

Electric water heating A system which uses an immersion heater installed into the hot storage vessel. This behaves in a similar way to the element in an electric kettle; when the desired temperature is achieved, sensed by a thermostat, the element is switched off. It is essential that the heater element reaches low down near the bottom of the storage vessel because it will only heat the water above the depth to which the element will reach. The heater should be at least 50 mm from the base of the storage vessel to prevent convection currents disturbing any sediment. Sometimes two heater elements are used, one fitted at low level and one much nearer the top. The higher immersion heater is switched on if, for example, only enough hot water is required to fill a sink, whereas should enough hot water to fill a bath be required, the lower heater is switched on.

The electrical power supply to an immersion heater must come directly from the consumer unit (see page 230) to terminate close to the hot storage vessel with a double pole switch. It is from here that a 21 amp heat-proof flex is run to the heater.

Gas storage heaters Purpose-made vessels which have gas burners installed directly below the stored water. The system incorporates an open flue which must be discharged to the external environment, the flue passing up through the storage cylinder.

Boiler–cylinder system A system in which a small boiler (e.g. a gas circulator) is located somewhere in the building and the hot water is conveyed via primary flow and return pipes. When using this system to heat the dhw it is not possible to include radiators on the primary pipework, unless they are made of non-ferrous metals. This is because the water which passes to the boiler is being supplied by the feed cistern and is constantly passing through the pipework as it is drawn off at the taps, thus bringing in entrapped gases which will cause atmospheric corrosion to ferrous metals. Direct systems which use primary flow and return pipes should not be used in hard water areas owing to the problems of scale build up in the pipework, blocking the flow.

Instantaneous systems It is possible to heat the water directly by passing it through a heat exchanger. Several systems are available, e.g. the Multi-point, the water-jacketed tube heater and the combination boiler. The main disadvantage with instantaneous heaters is the fact that only a limited number of draw-off points can be supplied at once, because of the restricted flow rate through the heat exchanger.

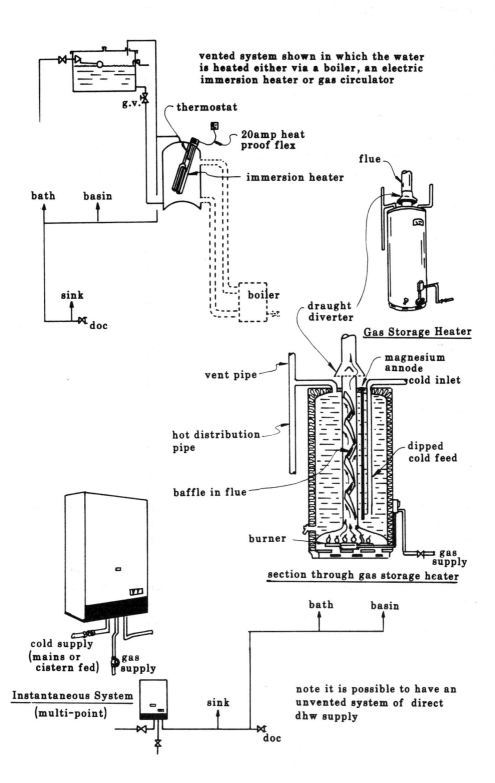

vented system shown in which the water
is heated either via a boiler, an electric
immersion heater or gas circulator

g.v.

thermostat

20amp heat
proof flex

immersion heater

flue

bath   basin

sink

doc

boiler

draught
diverter

Gas Storage Heater

vent pipe

magnesium
annode
cold inlet

hot distribution
pipe

dipped
cold feed

baffle in flue

burner

gas
supply

section through gas storage heater

bath   basin

cold supply
(mains or
cistern fed)

gas
supply

Instantaneous System
(multi-point)

sink

note it is possible to have an
unvented system of direct
dhw supply

doc

Direct Hot Water Supply

Indirect Hot Water Supply

Relevant British Standards
BS 5546 and BS 6700

The indirect system is probably the most common form of dhw, allowing a boiler to be used also for central heating purposes. The storage vessel is the heart of these systems and consists of a special cylinder in which is fitted a heat exchanger. The heat exchanger allows water from the boiler circulating pipework (primary pipework) to pass through, but not mix with, the water in the cylinder itself. Thus in effect it really consists of two systems which appear to join at the hot storage vessel.

Water is heated in the boiler and conveyed to the hot storage vessel via primary flow and return pipes by gravity circulation (convection currents) or by the use of a circulating pump. The water supplying this primary circuit can be taken from the f & e cistern, which is usually located in the roof space, or directly from the supply main in the case of a sealed system (see Part 3 Central Heating). The water, once in the primary pipework, is never changed, except for maintenance purposes; therefore any calcium carbonate (limescale) would have been precipitated and any gases which came in with the fresh supply would have escaped from the water. Thus, the water in this state is somewhat neutralised; it will not cause excessive corrosion and is suitable for central heating purposes as it will not corrode the steel radiators.

Because the primary water must not be changed, the water supply for domestic purposes needs to be taken from a separate supply, and, if using a feed cistern in the roof space, it must be separate from the f & e cistern to prevent the mixing of the waters in each system.

As we have seen, most indirect systems are of the double feed design, using two cisterns in the roof space, and having a coil or annulus type heat exchanger fitted into the cylinder. But there is a second type, known as a single feed indirect system (the design in the figure being known as the Primatic). This system uses a specially designed heat exchanger which allows the primary circuit to fill up via a built-in feed pipe; in so doing it maintains an air break separating the primary and secondary waters.

The figure shows how the expansion of the water in the primary circuit is taken up by moving the air in the top dome back through its cold feed pipe to the lower dome. The primary circulation system must not be too large (having lots of radiators) because the excessive quantity of water that the system will contain would, when expanding, exceed that of the space available in the dome, thus forcing the air out; also, a circulating pump must not be used on the primary circuit to the dhw cylinder for the same reason. Should the air be lost the space would be filled instead with water and it would be converted, in effect, into a direct system of hot water supply; this would give rise to corrosion problems. This type of system must not be pressurised because the mains pressure would blow out the air lock needed to keep the primary and secondary waters separate.

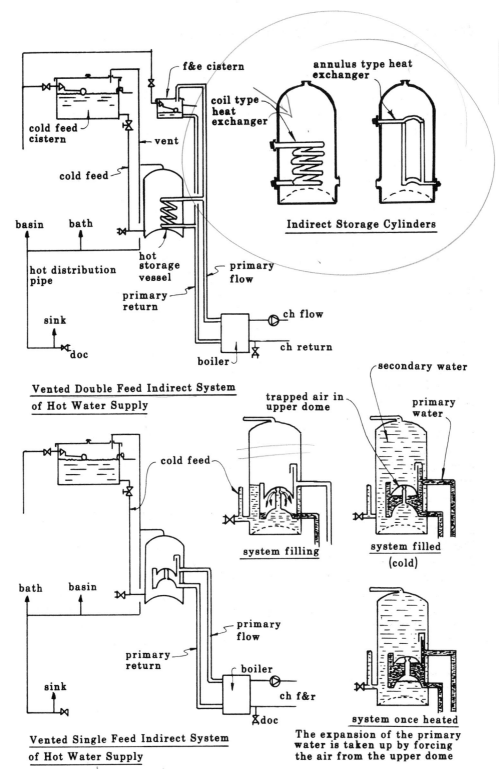

f&e cistern

annulus type heat exchanger

coil type heat exchanger

cold feed cistern

vent

cold feed

basin   **bath**

hot distribution pipe

hot storage vessel

primary flow

primary return

Indirect Storage Cylinders

sink

doc

ch flow

ch return

boiler

Vented Double Feed Indirect System of Hot Water Supply

cold feed

secondary water

primary water

trapped air in upper dome

system filling

system filled (cold)

bath   **basin**

primary flow

primary return

boiler

ch f&r

doc

sink

system once heated

Vented Single Feed Indirect System of Hot Water Supply

The expansion of the primary water is taken up by forcing the air from the upper dome

Indirect Hot Water Supply

Unvented Domestic Hot Water Supply

Relevant British Standard
BS 6700

When considering the installation of a system of unvented dhw (stored supply in excess of 15 litres) Part G of the Building Regulations should be observed; this identifies several requirements to be met by the local authority. Systems must be installed by an 'approved' installer who is registered with a recognised body and the system must be purchased as a *unit* or *package*.

A *unit* is a system in which all the component parts have been fitted by the manufacturer at the factory. A *package* is a system in which the temperature-activated controls are incorporated but all other components are fitted by the installer. In both cases this ensures that the safety devices, which are 'factory set', are installed with the system.

The regulations state that at no time must the water reach 100°C, which is ensured by the use of three safety devices: the thermostat (operating at 60°C); a high temperature thermal cut-out device (which locks out at 90°C); and a temperature relief valve (designed to open at 95°C).

With unvented systems the water is taken directly from the mains water supply. There is no open vent pipe or storage cistern where the expansion of heated water can be taken up; therefore a sealed expansion vessel is incorporated in the cold feed pipework. To comply with byelaw requirements a check-valve must be fitted on the supply pipe to prevent a backflow of hot water down the supply main. Should the expansion vessel not function for any reason the water, on expanding, will be forced out of the pressure relief valve.

Any water discharging from either a temperature or pressure relief valve must be conveyed, via an air break, to a suitable discharge position. The discharge pipe must not exceed 9 m in length and the pipe diameter of the discharge outlet must be maintained.

A pressure-reducing valve is fitted as a precaution to reduce excessive water pressures which may cause damage to the system. In order to ensure equal pressures at both hot and cold draw-off points, the cold supply pipe is sometimes branched off after this valve.

The advantages of these systems include the higher pressures obtainable at the draw-off points, the use of less pipework and the fact that less time is required for installation. The disadvantages are frequently overlooked. Firstly, such a system can only be installed should the flow rate (volume of water) be sufficient to supply both hot and cold water at once, bearing in mind several appliances could be running at the same time. Secondly, in hard water districts the build up of scale around temperature and pressure relief valves could make those valves ineffective. Regular servicing of the system is, therefore, essential.

Some unvented dhw cylinders incorporate an unvented dome which traps a pocket of air, thus doing away with the need for a sealed expansion vessel. Unfortunately high water turbulence within the cylinder causes the air trap to be lost, also after a period of time the air will be absorbed into the water causing its ineffective use, resulting in the pressure relief opening. Consequently the cylinder will need draining down to recharge the dome with air.

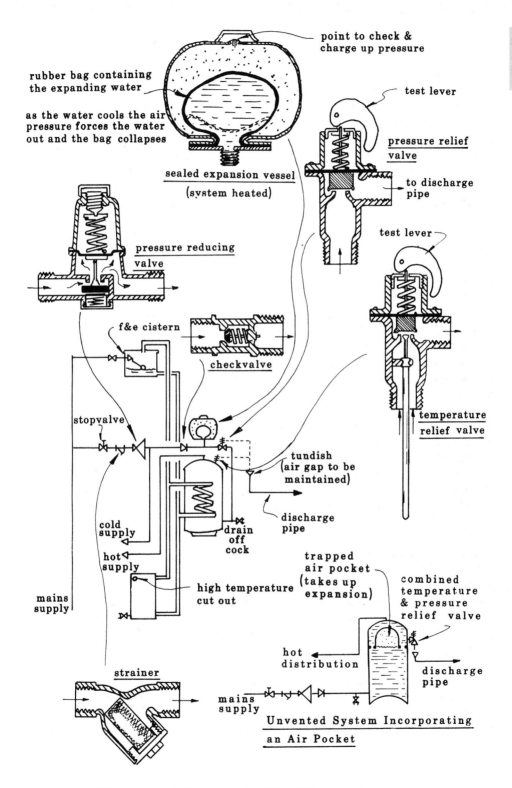

point to check &
charge up pressure

rubber bag containing
the expanding water

as the water cools the air
pressure forces the water
out and the bag collapses

sealed expansion vessel
(system heated)

test lever

pressure relief
valve

to discharge
pipe

pressure reducing
valve

test lever

f&e cistern

checkvalve

temperature
relief valve

stopvalve

tundish
(air gap to be
maintained)

cold
supply

hot
supply

discharge
pipe

drain
off
cock

mains
supply

high temperature
cut out

trapped
air pocket
(takes up
expansion)

combined
temperature
& pressure
relief valve

hot
distribution

discharge
pipe

strainer

mains
supply

Unvented System Incorporating
an Air Pocket

Unvented Domestic Hot Water Supply

Hot Distribution Pipework

Relevant British Standard
BS 6700

The hot water in the storage vessel needs to be preserved for as long as possible to save on fuel consumption; therefore the cylinder and pipework should be insulated. However, hot water can also be lost due to one pipe circulation, which is the circulatory flow of hot water, by convection currents, up the vertical vent pipe. To prevent this the hot distribution should be run a minimum of 450 mm horizontally on leaving the top of the hot storage vessel.

When a tap is opened before the hot water can discharge out from the spout the cold water in the pipe has to be drawn off and is invariably allowed to run to waste; also there is a certain amount of inconvenience to the user. This run of pipe to the appliance is referred to as a 'dead leg' and should not exceed that indicated in the table.

Maximum length of uninsulated hot distribution pipe

Internal bore of pipe (mm)	Length (m)
less than 10	20
11–19	12
20–25	8
over 26	3

Where it is not possible to keep within these limits the pipe should be thermally insulated, or some other method should be used to ensure that the hot water appears quickly at the tap. Either of two methods can be adopted to meet this requirement: a specially designed heat tracing tape can be used, which heats up as necessary, maintaining the water temperature; or a system of secondary circulation will need to be installed.

Secondary circulation An arrangement in which a pipe is run back to the dhw cylinder from the furthest point on the distribution, thus forming a circuit. Water can flow around this circuit either by convection currents or by the use of a non-corrosive circulating pump, thus allowing hot water to be kept close to the draw-off points. The return pipe is connected within the top third of the cylinder to prevent the cooler water, lower down the cylinder, mixing with the hot and reducing its temperature.

Hydraulic gradient When a tap is opened and water drawn from the system, the water level in any vertical pipe will drop; the amount by which it drops will depend on the size of the pipework and the flow rate being drawn off (see figure). In any cistern-fed system, be it hot or cold, if the pipework is less than perfect when several appliances are running at once, one will be starved of water (usually the highest draw-off point); this is due to the above-mentioned drop in water level in the vertical pipe.

When installing a system of hot water supply it is particularly important to consider this design concept, because any high connections in the vertical rise of the vent pipe will be starved of water unnecessarily.

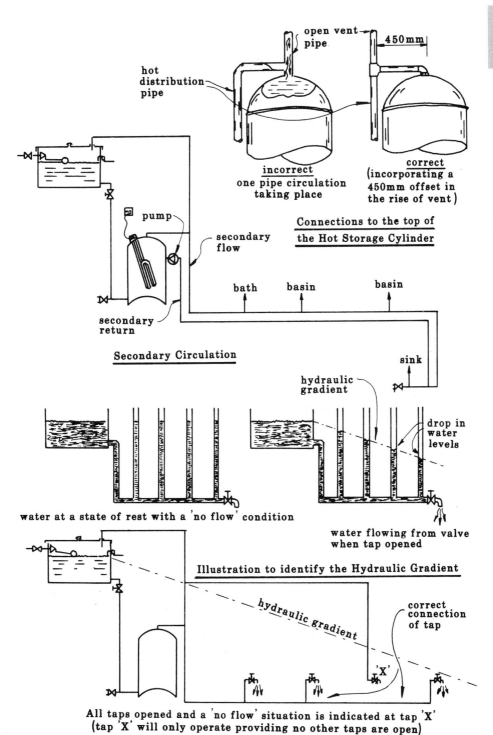

open vent
pipe

450mm

hot
distribution
pipe

incorrect
one pipe circulation
taking place

correct
(incorporating a
450mm offset in
the rise of vent)

Connections to the top of
the Hot Storage Cylinder

pump

secondary
flow

bath   basin   basin

secondary
return

sink

Secondary Circulation

hydraulic
gradient

drop in
water
levels

water at a state of rest with a 'no flow' condition

water flowing from valve
when tap opened

Illustration to identify the Hydraulic Gradient

hydraulic gradient

correct
connection
of tap

'X'

All taps opened and a 'no flow' situation is indicated at tap 'X'
(tap 'X' will only operate providing no other taps are open)

Hot Distribution Pipework

Heat Recovery Period

This is the amount of time required to heat up a quantity of water. The recovery time varies, depending upon the power rating of the heat source used, but basically the higher the power rating the shorter the heating up period of the water. It is possible to calculate the heat recovery period using the following formula:

$$\frac{SH \times kg \times \text{temperature rise}}{\text{available time in seconds}}$$

SH = specific heat of water 4.186 kJ
kg = the weight of water to be heated (i.e. 1 litre = 1 kg)

Example: Find the power required to heat 100 litres of water in 2½ h from 4°C to 60°C.

Therefore $\dfrac{SH \times kg \times \text{temp rise}}{\text{seconds in 2½ h}} = \dfrac{4.186 \times 100 \times 56}{9000} = 2.6 \text{ kW}$

Note that the above example ignores heat loss and the heating up of the hot storage vessel itself; it would be advisable to add, say, 10% for this; thus 2.6 + 10% = 2.86 kW, and as a result a 3 kW heater would be chosen.

It will be seen that the incoming temperature of the water has a bearing upon the recovery time; therefore, one can only estimate. Often the householder will require some guidance concerning the time it will take to heat the water, given an immersion heater already supplied. This is found using the above formula, although it will need transposing to suit this new situation.

Example: Find out how long it will take to increase the temperature of 136 litres of water from 10°C to 60°C, using a 3 kW immersion heater. Ignore heat losses, etc.

$$\text{formula} = \frac{SH \times kg \times \text{temp rise}}{\text{time in seconds}} = \text{power}$$

$$\text{transpose to } \frac{SH \times kg \times \text{temp rise}}{\text{power (kW)}} = \text{time in seconds}$$

Therefore $\dfrac{4.186 \times 136 \times 50}{3} = 9488 \text{ s or } 2.6 \text{ h (2 h 36 min)}$

Sometimes a recovery chart is used to give an approximate time and proves a useful guide when sizing up small water heaters or boilers,

Recovery chart: to heat water through 55°C

Litres	Power rating (kW)									
	1	2	3	4	6	8	10	12	15	18
5	19	10	6	5	3	2	2	2	1	1
10	38	19	13	10	6	5	4	3	3	2
15	58	29	19	14	10	7	6	5	4	3
25	96	48	32	24	16	12	10	8	6	5
50	192	96	64	48	32	24	19	16	13	11
100	384	192	128	96	64	48	38	32	26	21
150	576	288	192	144	96	72	58	48	38	32
200	767	384	256	192	128	96	77	64	51	43

Time in minutes

Example: To heat 150 litres of water from 5°C to 60°C in 144 minutes a boiler with a rated boiler output of 4 kW would be required.

Where the water is heated in a boiler, etc., away from the hot storage vessel, it will need to be conveyed via primary flow and return pipework. Circulating this water to and from the storage vessel can be achieved by gravity circulation or speeded up by the use of a circulating pump (see page 122, Fully Pumped System). The greater the circulation pressure from the heat source to the storage vessel, the quicker will be the heat recovery period. Where gravity circulation is chosen to convey the hot water it should be understood that the greater the circulating head the greater will be the circulating pressure. Where only a poor- or low-circulating head can be achieved, larger pipe sizes will be required, or the inclusion of a circulating pump.

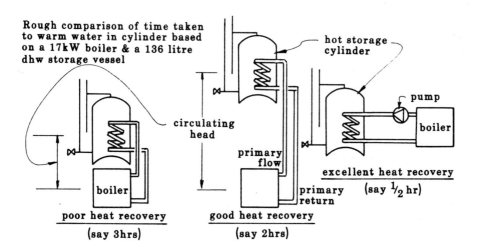

Heat Recovery Period

Instantaneous Domestic Hot Water Supplies (Centralised)

Relevant British Standards
BS 5546 and BS 6700

With instantaneous dhw systems the principle is to pass the cold water through a heat exchanger, such as a coil of pipework passing through a heat source, to which the water is heated by the time it comes out the other end. There is a limit to the speed at which the water can be heated; therefore the flow rate (volume) of water needs to be minimised; failure to minimise the flow rate will result in an insufficient heat-up. Because of this reduced flow rate of water passing through the heat exchanger it will not be possible to supply several outlets at once; as a result these systems are unsuitable where there is to be high demand on hot water.

The advantages of instantaneous heaters include the fact that the water pressure can be maintained, if supplied from the supply main, to provide suitable shower facilities, etc., and because no storage of dhw is provided the Building Regulations are not directly applicable. With these systems you only heat the water as and when it is required; therefore, a saving can be made in fuel consumption. Instantaneous heaters include multi-points, water-jacketed tube heaters and combination boilers. (See page 128 for combination boilers.)

The Multi-point

This consists of a gas burner located beneath a heat exchanger. When hot water is required the water, in passing through the heater, causes the gas valve to open, and this is ignited by a pilot flame. The gas valve opens as the result of a reduction in the water pressure on one side of a diaphragm located in the pressure differential valve. This pressure reduction is caused by water passing through a Venturi which creates a negative pressure sucking the water from the valve. Attached to the diaphragm is the push rod which opens the gas line. On shutting the water supply the pressure in the differential valve equalises and the diaphragm is sprung shut.

The water-jacketed tube heater

Sometimes referred to as a thermal storage system, this is a system in which the water to be heated is passed through a stored supply of central heating water. It is like an indirect system of dhw in reverse; the domestic water flows through the heat exchanger and not the primary water. One end is connected to the cold supply main, the other directly to the taps. When the water is heated a small amount of expansion occurs and is taken up in a small expansion vessel located in the cold supply line fitted to the unit or within the unit itself via an expansion chamber. Because the water in the heat exchanger can potentially become as hot as the water in the primary storage cylinder, it must be noted that water, upon leaving the unit, passes through a thermostatic mixing valve in which the water is cooled from the cold supply to give a temperature no hotter than 60°C.

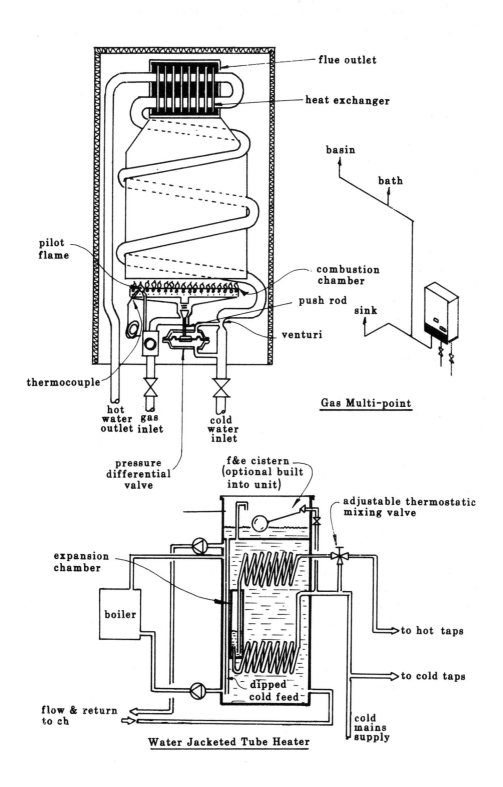

flue outlet

heat exchanger

basin

bath

pilot flame

combustion chamber

push rod

sink

venturi

thermocouple

hot water outlet | gas inlet | cold water inlet

Gas Multi-point

pressure differential valve

f&e cistern (optional built into unit)

adjustable thermostatic mixing valve

expansion chamber

boiler

to hot taps

to cold taps

dipped cold feed

flow & return to ch

cold mains supply

Water Jacketed Tube Heater

Instantaneous Domestic Hot Water Supplies

Localised Hot Water Heaters

Relevant British Standards
BS 5546 and BS 6700

Two distinct types of localised dhw heaters will be found: the instantaneous and storage types. In each case the heater will only serve one sink – two if fitted in close proximity to the heater.

Instantaneous single points

These heaters are fuelled either by gas or electricity and heat the water only when required. They are usually fitted with a swivel spout and located directly above the sanitary appliance, the water flow usually being inlet-controlled. The gas heater works on the same principle as the Multi-point (see page 88).

With electric instantaneous heaters water is allowed to flow into the heater, where it is surrounded by an electric heating element. Because of the small volume of water surrounding the element the water quickly heats up as it is drawn through the heater. The temperature of the water will be directly related to the kW rating of the appliance and the water flow rate. The water flowing through the heater is sensed by the pressure or flow switch located on the inlet supply, which in turn makes the electrical contacts to the immersion heater element.

Storage type single point

These heaters are located either above or below a sink or similar appliance and have a capacity not exceeding 15 litres. The stored water is heated by an electric element to, say, 60°, and on expanding the water is allowed to push up and discharge out from the discharge spout. It is important to make the client aware of this dripping spout. When cold water enters the base of the unit it forces the hot stored water out. Obviously the discharge of hot water is limited and will soon start to cool, but it will be sufficient for small quantities of draw-off. When installing these heaters below an appliance a special design of terminal fitting (tap) needs to be used – one which allows water to flow through the heater but at the same time allows the water to expand (see figure).

Some designs of single points incorporate a small expansion vessel which enables an outlet control valve to be used, eliminating dripping outlet spouts.

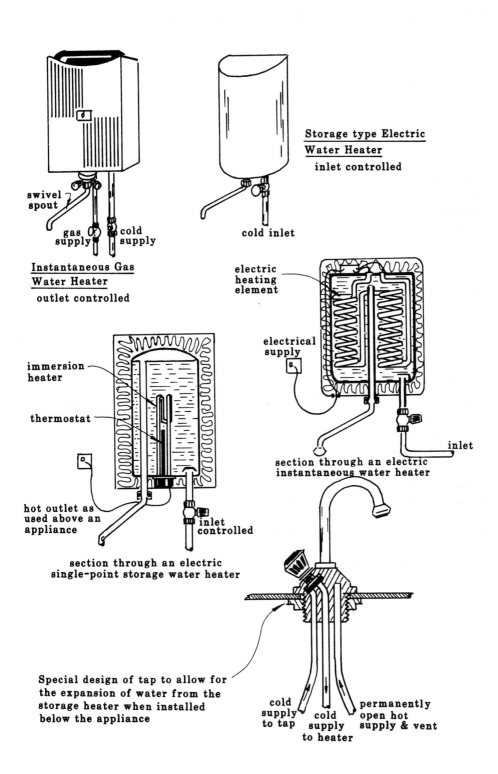

swivel
spout

gas
supply   cold
             supply

**Instantaneous Gas
Water Heater**
outlet controlled

Storage type Electric
Water Heater
inlet controlled

cold inlet

electric
heating
element

electrical
supply

immersion
heater

thermostat

hot outlet as
used above an
appliance   inlet
                controlled

section through an electric
single-point storage water heater

section through an electric
instantaneous water heater

inlet

Special design of tap to allow for
the expansion of water from the
storage heater when installed
below the appliance

cold
supply   permanently
to tap   cold   open hot
         supply supply & vent
         to heater

Localised Hot Water Heaters

Solar Hot Water System

Relevant British Standard
BS 5918

This is a dhw system which utilises solar energy from the sun. It collects the radiant heat waves in a solar collector usually located on the roof. A simple solar collector consists of a thin vessel painted matt black with piping attached flowing to and from a hot storage vessel. It is covered with double- or triple-glazing and backed by thermal insulation material. The collector is sited at a convenient position to catch the solar heat, usually at an angle of 40° and facing south.

Designs of solar heating can vary, and because of the unreliable weather in Great Britain, it would not be used in any dwelling as the only form of heating for water. Generally cold water is supplied to a warm store vessel which is heated by solar energy; the water then passes onto the normal conventional hot storage cylinder and is supplied with additional heat (if required) via a boiler or electric immersion heater. Note that in the system shown the solar collector is fitted above the cold feed cistern; this causes no problems because the primary circuit from the solar collector forms part of a closed system. The water in the primary circuit is made to circulate by means of a pump which is switched on automatically should the temperature in the top of the collector be higher than that in the base of the hot storage vessel. However, it is possible to design a gravity system, providing the collector can be located sufficiently below the warm store vessel to allow circulation to take place.

Thermal performance Based on a long-term average temperature in London, the system should supply an approximate percentage of that indicated in the following table. However, the amount of solar energy supplied to the dhw system will vary from area to area, and will depend upon the effectiveness of the solar collector and its location. It should be remembered that trees and buildings causing shade will significantly reduce system performance.

Approximate monthly percentage for solar heating system, in the London area, on a south facing 30° pitched roof

	%		%
January	2	July	13
February	5	August	13
March	6	September	12
April	10	October	8
May	12	November	4
June	13	December	2

Design considerations Due to temperature changes the temperature of the heat transfer liquid could vary from about −15°C to as high as 200°C, when not circulating. Where water is used for this medium (see figure for example of this) it will be necessary to prevent the temperature from rising above 100°C by incorporating a pressure relief valve as a minimum requirement, discharging its contents to a safe location. To prevent damage due to excessively cold conditions an anti-freeze solution may be added; alternatively the system will need to be drained in winter. Usually planning permission will need to given by the local planning authority before a solar heating system can be installed.

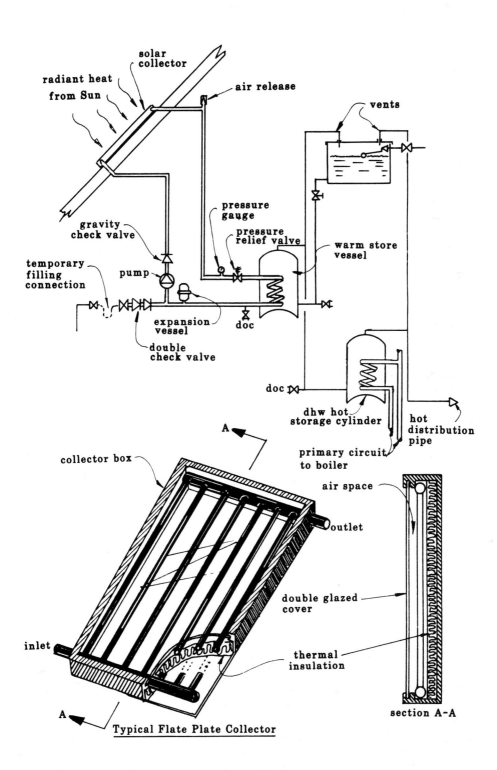

Solar Hot Water System

Connections to Hot and Cold Pipework

Relevant British Standard
BS 6700

Connections to showers

The hot and cold supplies to a shower will need to be of equal pressure. Where the supply is directly from the service main, provision must be maintained to ensure no backflow occurs. This is usually achieved by ensuring that the shower head cannot discharge below the overspill level of the appliance or by incorporating a double check-valve assembly into the pipeline.

Connections via a storage cistern will need to be such that an adequate pressure is achieved; in general a minimum distance of 1 metre from the underside of the cistern to the shower head should be maintained. The pressure can be increased by the use of a booster pump fitted into the pipeline in which a small self-contained unit, designed to give a greater head of water, is used. The only proviso is that at least 150 mm of initial head is available to allow the flow switch to operate and start the pump when the supply is opened.

It is possible to use a flow-activating button where no head at all is available for some designs of pump. Booster pumps may be fitted prior to, or after, the shower-mixing valve although in general they should be installed in such a location as to ensure that it is constantly flooded with water.

To ensure the shower is never starved of water the cold supply to the shower mixing valve should be independent of other draw-off points and its connection to the storage cistern below that of the cold feed to the dhw cylinder.

The temperature of the water to the shower is regulated by means of manual or thermostatic control. With the manually controlled valve a dial on the control head is turned to open or close either the hot or cold port hole size, thus restricting the flow. Thermostatic mixing valves are fitted with a temperature-sensing device which is designed to expand due to heat and should maintain a constant outlet temperature, opening or closing the port holes respectively and automatically.

Connections to bidets

There are two types of bidet: those with pillar taps to give an over-the-rim type discharge, thus maintaining an air gap; and those with a submerged nozzle which discharges a spray of water upwards from the base of the appliance. Those with an ascending spray are not permitted to be connected directly to the supply main and must have their hot and cold supplies run via separate distribution pipes, independent of other draw-offs. This can be achieved in the way shown in the figure. Note that a check valve and an additional separate vent pipe are required from the hot distribution pipe to the appliance.

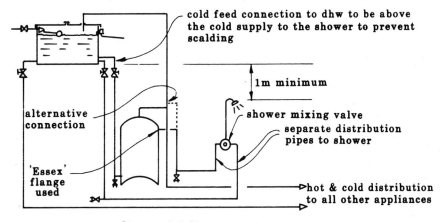

cold feed connection to dhw to be above the cold supply to the shower to prevent scalding

1m minimum

alternative connection

shower mixing valve

separate distribution pipes to shower

'Essex' flange used

hot & cold distribution to all other appliances

Storage fed Shower

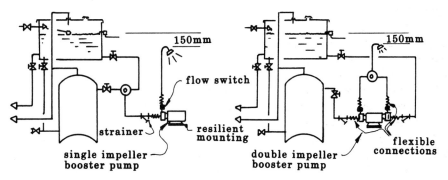

150mm

flow switch

strainer

resilient mounting

single impeller booster pump

150mm

flexible connections

double impeller booster pump

Installation of Shower Booster Pumps

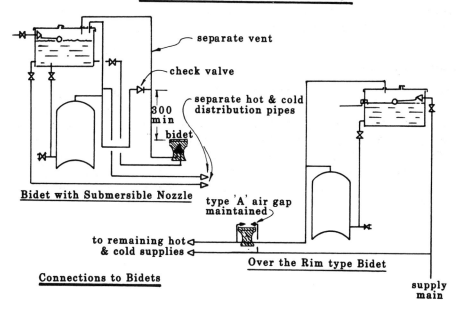

separate vent

check valve

separate hot & cold distribution pipes

300 min

bidet

Bidet with Submersible Nozzle

type 'A' air gap maintained

to remaining hot & cold supplies

Over the Rim type Bidet

supply main

Connections to Bidets

Connections to Hot and Cold Pipework

Installation of Pipework 1

Relevant British Standard
BS 6700

Pipe supports

There are many designs of pipe support bracket and the one chosen will depend upon the material nature of the pipe and the cost allowed for the job. The latter would in turn depend upon circumstances; for example, it would be pointless to use plastic pipe clips in schools or hospitals, etc., where they could very easily be damaged. Whatever pipe support is chosen the fixing must be secure to prevent damage and the possible development of air locks. As a guide the general recommended pipe support spacings are given in the following table, but one must remember that one clip too many is better than one clip too few and in many cases plumbers have to use their own judgement.

Maximum spacings for internal pipework (m)

Pipe size		Copper pipe		Steel pipe		Plastic pipe	
(mm)	(in)	(horiz)	(vert)	(horiz)	(vert)	(horiz)	(vert)
15	½	1.2	1.8	1.8	2.4	0.6	1.2
22	¾	1.8	2.4	2.4	3.0	0.7	1.4
28	1	1.8	2.4	2.4	3.0	0.8	1.5
35	1¼	2.4	3.0	2.7	3.0	0.8	1.7
42	1½	2.4	3.0	3.0	3.6	0.9	1.8
54	2	2.7	3.0	3.0	3.6	1.0	2.1

Design considerations

If the pipe is to run through structural timbers such as floor joists, it is essential that the structural members are not weakened. Notches and holes should be as small as practicable, but should also allow for pipe expansion and contraction. The size and position of a notch or hole needs to be considered and should not exceed the dimensions in the figure.

Example: For a joist 200 mm deep and 3 m long any notch must have a maximum depth of $H \div 8$

$$\text{Therefore } 200 \div 8 = 25 \text{ mm}$$

and be at least $7L \div 100$ from its bearing. Therefore $7 \times 3000 \div 100 = 210$ mm

and no greater than $L \div 4$ from its bearing. Therefore $3000 \div 4 = 750$ mm

In vented systems the pressure is usually quite poor in comparison with mains supply pipework; therefore it is essential to run the pipework with no dips or high spots, which may allow a trap of air to form, causing a blockage (air lock). To this end pipes should be run horizontal, or to an appropriate fall, allowing the air to escape from the system.

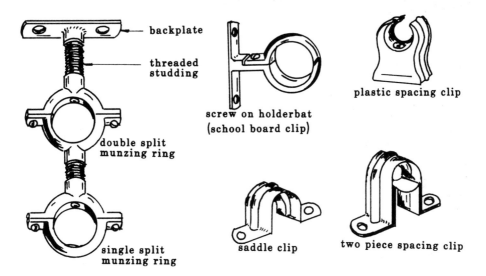

backplate

threaded
studding

double split
munzing ring

single split
munzing ring

screw on holderbat
(school board clip)

plastic spacing clip

saddle clip

two piece spacing clip

Typical Pipe Supports

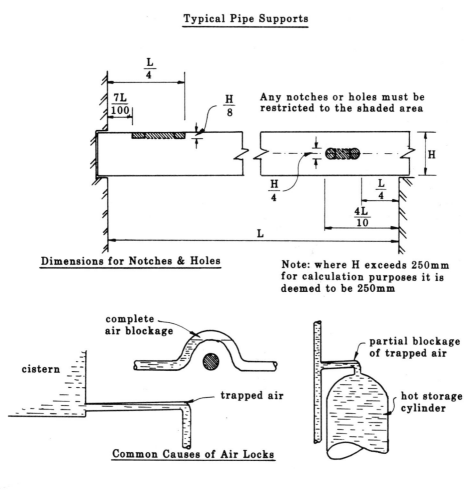

$\frac{L}{4}$

$\frac{7L}{100}$

$\frac{H}{8}$

Any notches or holes must be
restricted to the shaded area

$\frac{H}{4}$

$\frac{L}{4}$

$\frac{4L}{10}$

H

L

Dimensions for Notches & Holes

Note: where H exceeds 250mm
for calculation purposes it is
deemed to be 250mm

complete
air blockage

cistern

partial blockage
of trapped air

trapped air

hot storage
cylinder

Common Causes of Air Locks

Installation of Pipework 1

Installation of Pipework 2

Relevant British Standards
BS 5422 and BS 6700

Thermal insulation

When installing any pipework, steps need to be taken to prevent the pipe contents from *freezing* in the case of cold water supplies and *losing heat* in hot water systems. Therefore, some form of insulation may be required. Insulation materials entrap small air pockets, still air being a poor conductor of heat. Should a pipe freeze, damage can occur to the pipework; therefore insulation is imperative in exposed locations. Roof spaces and garages must be regarded as exposed. The insulation requirements depend on the thermal conductivity of the insulation material. The following table gives a guide to the minimum insulation for frost protection in housing.

Minimum thickness of insulation material

Outside pipe diameter (mm)	Indoor installations (e.g. roof space)		Outdoor installations (including below ground)	
	Flexible foam and expanded plastic (mm)	Loose-fill (mm)	Flexible foam and expanded plastic (mm)	Loose-fill (mm)
0–15	22	89	27	100
15–22	22	75	27	100
22–42	22	75	27	89
42–54	16	63	19	75
Flat surfaces	13	38	16	50

It should be noted that smaller pipes require a greater thickness of material because they are more prone to freezing up.

Accessibility of pipework

It is a water byelaw requirement that all water pipes and fittings are readily accessible for inspection and repair. Often the designer/installer has no wish to see exposed pipework; therefore, a duct or chase is made in the wall, etc., and enclosed by a cover to allow movement and ease of removal. This cover may be covered at the choice of the client with plaster or a tile finish (see figure for examples). Note that thermal insulation material may also be required in exposed or unheated dwellings, although not shown in the examples. It is only possible to encase the pipe in the floor screed where it forms part of a heating element of a closed circuit of under-floor radiant heating.

When passing pipes through floors or walls the pipe should be sleeved to allow movement, and the space between the pipe and sleeve 'fire stopped' to prevent the passage of smoke and flame.

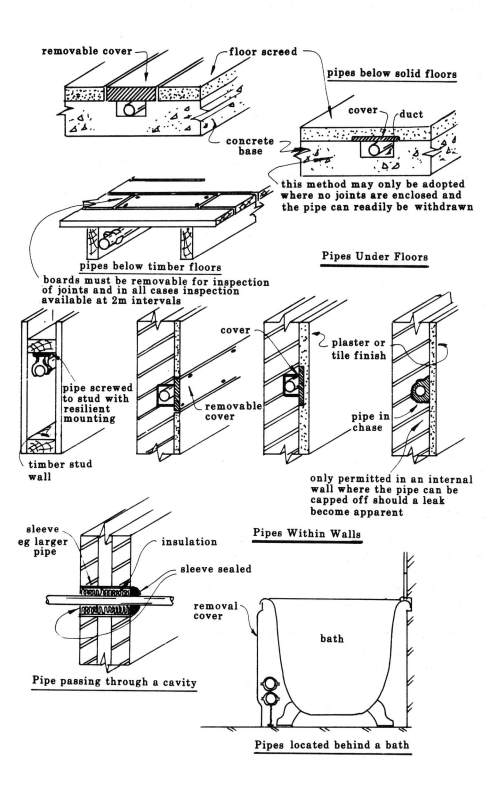

removable cover — floor screed

pipes below solid floors

cover — duct

concrete base

this method may only be adopted where no joints are enclosed and the pipe can readily be withdrawn

pipes below timber floors

Pipes Under Floors

boards must be removable for inspection of joints and in all cases inspection available at 2m intervals

pipe screwed to stud with resilient mounting

cover

plaster or tile finish

removable cover

pipe in chase

timber stud wall

only permitted in an internal wall where the pipe can be capped off should a leak become apparent

Pipes Within Walls

sleeve eg larger pipe

insulation

sleeve sealed

removal cover

bath

Pipe passing through a cavity

Pipes located behind a bath

Installation of Pipework 2

Pipe Sizing of Hot and Cold Pipework

Relevant British Standard
BS 6700

For everyday plumbing in a domestic dwelling the installer uses a rule of thumb (tried and tested) method which consists of 15–22 mm pipe on the supply main and a 22–28 mm pipe for the hot and cold distribution pipework; in each case the larger size is chosen if there are many draw-off points, the pipe slowly reducing in size as necessary to each appliance. For bigger systems requiring many outlets over a large area or several floors the main distribution pipe run will need to be sized correctly to ensure sufficient pressure and flow at the draw-off points, without excessive noise problems.

Shown opposite is a completed example, which acts as a key to the figure below it. An explanation of the stages carried out to confirm the pipe sizes chosen is as follows.

Column 1 This is the pipework which is being sized; note that the system is broken into various sections.

Column 2 The flow required is found by making an assessment of the probable maximum demand of water, in litres per second, at any given time, because it is very unlikely that all the sanitary appliances will be used at once. To perform this assessment a method has been devised based upon the theory of probability in which a loading unit rating is given to each type of sanitary appliance.

Sanitary appliance	Loading unit
WC cistern	2
Bath	10
Wash basin	1½
Sink and washing machine	3
Shower	3
Bidet	1

By multiplying the number of each type of appliance by its loading unit and adding all the results the total loading units for the system will be found.

To convert loading units to litres per second the conversion table opposite is used. Thus, in our example each floor has one bath, two basins and two WC cisterns.

$$
\begin{aligned}
\text{Therefore:} \quad 1 \times 10 &= 10 \\
2 \times 1\tfrac{1}{2} &= 3 \quad + \\
2 \times 2 &= 4 \\
\hline
&\underline{17} \text{ loading units}
\end{aligned}
$$

This, compared to the chart, gives 0.4 litres/s. Note that for pipe section A all five floors are being served; therefore the total loading units for this section will be: $17 \times 5 = 85$, which converts to 1.1 litres/s.

Cold distribution pipe serving five flats

1	2	3	4	5	6	7	8	9	
Section	Flow rate	Suggested pipe size	Velocity	Loss of head	Effective pipe length	Frictional head	Progressive head	Actual head	Notes
	(l/s)	(mm)	(m/s)	(m)	(m)	(m)	(m)	(m)	
A	1.1	28	1.8	0.2 × 16.5 =		3.3	3.3	3	Undersized
A	1.1	35	1.25	0.07 × 18.2 =		1.27	1.27	3	✔
B	0.92	28	1.5	0.15 × 4.5 =		0.68	1.95	6	✔
C	0.8	22	2.2	0.43 × 4.0 =		1.72	3.67	9	Possible noise
C	0.8	28	1.3	0.12 × 4.5 =		0.54	2.49	9	✔
D	0.6	22	1.6	0.27 × 4.0 =		1.08	3.57	12	✔

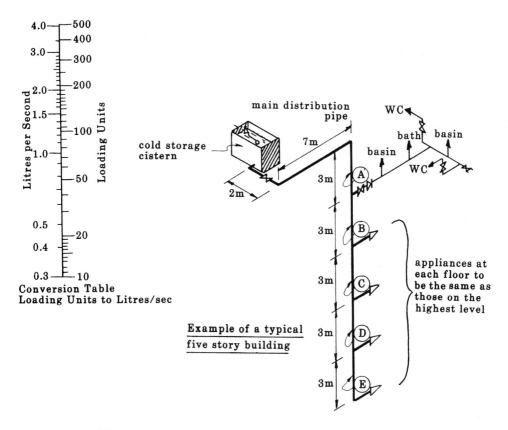

Conversion Table
Loading Units to Litres/sec

Example of a typical
five story building

Pipe Sizing of Hot and Cold Pipework

Column 3 An assumption is made at this stage that a particular pipe size is correct, and the table will confirm (or not) its possible use.

Columns 4 and 5 The velocity and loss of head are simply read from the graph opposite. To use, a horizontal line is taken from the flow in litres per second to intersect the pipe diameter; from this point the readings can be taken.

Column 6 The effective length of pipe run is found by adding the actual net length of pipe to the length of pipe due to frictional loss.

Equivalent pipe lengths, due to frictional loss through copper fittings

o.d. (mm)	Elbow (m)	Tee (m)	Stopcock (m)	Check-valve (m)
15	0.5	0.6	4.0	2.5
22	0.8	1.0	7.0	4.3
28	1.0	1.5	10.0	5.6
35	1.4	2.0	13.0	6.0
42	1.7	2.5	16.0	7.9
54	2.3	3.5	22.0	11.5

In the example given the effective length for section A, with an assumed pipe diameter of 28 mm, is:

$$\begin{array}{rl} \text{actual length} = & 12.0 \text{ m} \\ \text{three elbows} = & 3.0 \text{ m} + \\ \text{one tee} = & \underline{1.5 \text{ m}} \\ & 16.5 \text{ m} \end{array}$$

Column 7 The frictional head is found by multiplying the loss of head by the effective pipe length (i.e. column 5 × 6).

Column 8 The progressive head is the sum total of the frictional heads for each section above the section in question.

Example: The progressive head for section C will be sections A + B + C

Therefore   1.27 + 0.7 + 0.54 = 2.51 m

Column 9 The actual head is the total head available, measured vertically from the underside of the storage cistern to the end of the section of pipe in question.

In conclusion, one estimates a suggested pipe diameter and completes the table for the section in question to prove its suitability for use or not. The pipe size is correct, providing the progressive head does not exceed the actual head and the velocity does not exceed 2 m/s in cold water pipework and 1.5 m/s in dhw systems, thus limiting noise transmissions.

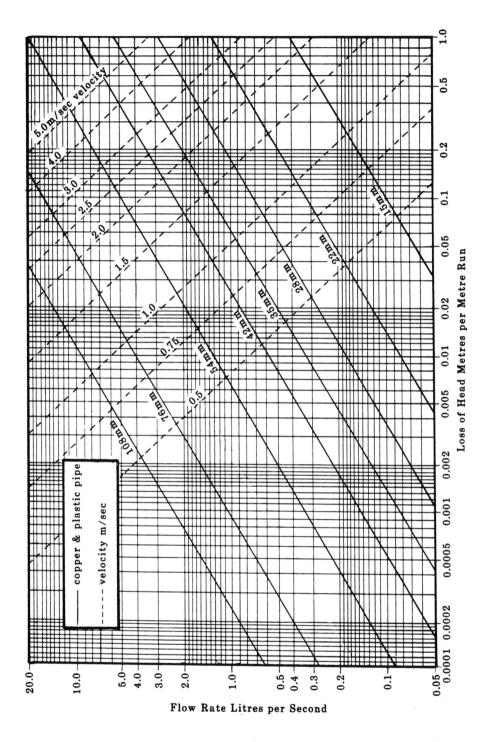

Pipe Sizing of Hot and Cold Pipework

(Reproduced with permission from *The Institute of Plumbing Design Guide*, The Institute of Plumbing, Hornchurch.)

Noise Transmission in Pipework

Relevant British Standard
BS 6700

Impulsive noise (water hammer) A hammering noise which occurs in high pressure water pipes; the noise is caused by surges of pressure. There are two basic noise types:

(1) A noise which consists of a sudden loud bang and is often caused by a loose stop-cock jumper or washer which quickly flips shut onto the seating; or it could be caused by pipes which have not been fixed correctly and flap about; this noise is caused by any sudden back surges of pressure, which could be created by the rapid closing of a tap.

(2) Oscillation or ballvalve murmur which consists of a series of bangs or rumbles generated in the pipeline. The noise is created by a float-operated valve quickly opening or closing, this being caused by ripples or waves which form on the surface of the water in a storage cistern. It is these ripples that cause the ball-float to bounce up and down, opening and closing the valve. To overcome this problem a larger ball-float is often used or a damping plate fitted to the float or ballvalve lever arm; alternatively, baffle vanes are fitted in the cistern to prevent waves forming.

One method often employed to cure water hammer is to shut down slightly the in-coming stop-valve; this does not reduce the pressure but it reduces the velocity and flow rate of the water.

Flow noise A general noise caused by the water flowing through the pipework. If the water velocity is kept below 3 m/s in high-pressure pipework and 2 m/s in low-pressure pipework this flow noise will not be significant. Sudden changes of direction and minimal downstream pressures can cause cavitation to occur, which is basically the swirling up of water flow, causing air bubbles to form and collapse. Flow noise can result unnecessarily when one uses roller pipe cutters and fails to remove the internal burr.

Pump noise When a booster pump has been provided noises should not be generated by water pressure and flows, providing the pump is correctly sized. However, the noise of the motor running may give concern, in which case isolation of the pump from the building is the only answer. Noise transmission from pumps can be reduced by using rubber-type connections to the pipework and mounting the pump on a resilient mounting.

In large buildings excessive water noise problems may be overcome by the installation of a hydropneumatic accumulator. This consists of a rubber bag in which the water can enter; the air surrounding the bag is charged to just below the system working pressure. Should a shock wave occur in the pipework the pressure surge is taken up by the cushion of air.

Splashing noise When water drops into cisterns the falling water can be somewhat noisy. To prevent this a collapsible silencer tube can be used, consisting of a polythene bag. It is sometimes possible to discharge the water onto an inclined plate, which in effect breaks the waterfall and inhibits the resulting splashing sound.

Movement noise When pipes, especially hot pipes, expand and contract they need to move and as a result must not be restricted (see pages 96–99 which identify good pipework installation practices).

Entrapped air bubbles and boiler noise This is the result of poor installation of pipework allowing air to be entrapped in dhw systems. Sometimes, owing to scale build-up in the flow from a boiler, a boiling noise is generated (called kettling) which is caused by steam forming and condensing. A similar noise is also caused where a flame impinges onto the heat exchanger, causing local hot spots.

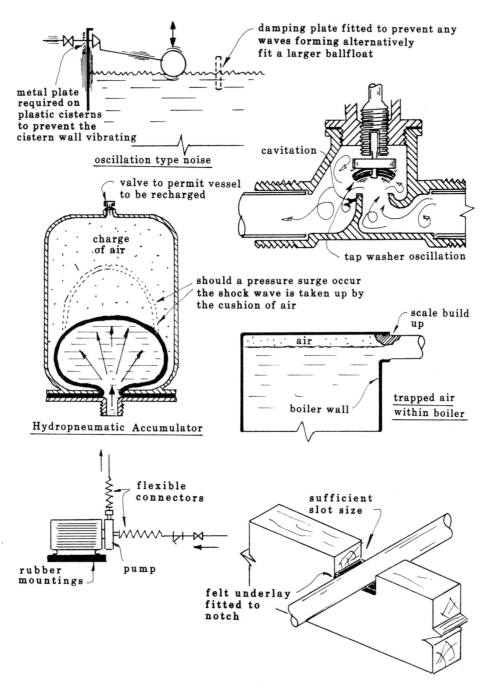

damping plate fitted to prevent any waves forming alternatively fit a larger ballfloat

metal plate required on plastic cisterns to prevent the cistern wall vibrating

oscillation type noise

cavitation

valve to permit vessel to be recharged

charge of air

tap washer oscillation

should a pressure surge occur the shock wave is taken up by the cushion of air

scale build up

air

Hydropneumatic Accumulator

boiler wall

trapped air within boiler

flexible connectors

sufficient slot size

rubber mountings

pump

felt underlay fitted to notch

Noise Transmission in Pipework

Commissioning of Hot and Cold Supplies

Relevant British Standard
BS 6700

When a system of hot or cold water supply has been installed the system should be inspected and tested as appropriate; the following checklist should be followed:

(1) **Visual Inspection** All pipework should be inspected to ensure it is fully supported and free from jointing compounds, flux, etc. The feed or storage cistern should be cleaned and made free of swarf. All valves should be closed to allow filling up in stages.

(2) **Test for Leaks** The system should be slowly filled, in stages; air should be expelled by opening the highest draw-off point on the section being tested. The testing is carried out in stages so that any leaks can be identified easily. Before opening the isolation valve to a cistern the float-operated valve seating should be temporarily removed to allow any grit, etc., in the pipe to flow into the cistern, thus preventing the blockage of the small hole in the seating itself. When the valve is reassembled the water level should be adjusted, as necessary.

Sometimes it is desirable to test the installation using a hydraulic test pump, giving a test pressure of one and a half to two times the system working pressure. This is achieved in the way shown in the figure, the test pressure being maintained for at least 1 hour. Pipework which is encased, prior to the connection of the water supply, must be tested in this fashion to prevent costly removal of the encasement cover.

(3) **System Flushing and Disinfection** All systems, large or small, will require the flushing through of the pipeline to remove flux residuals, wire wool, etc., from inside the pipe – usually achieved by opening the tap or draining off for a period of time. In the case of dhw the system should be flushed, both cold and hot; this will give a better removal of the deposits from within the system. The BS requires that all systems other than those of private domestic dwellings will need to be disinfected before being put into use. This is carried out after any initial flushing and is achieved by dosing the system with a measured quantity of sodium hypochlorite solution. The procedure outlined in BS 6700 (13.9) should be observed.

(4) **Performance Tests** Every terminal fitting, e.g. draw-off point, float-operated valve, should be checked for a suitable flow rate (volume of water) and pressure, and that it is thus operating to give suitable performance. For example, it would be pointless having an appliance with 3 bar pressure if the water only passed through a pin hole; it would take for ever to fill.

The performance test must be maintained under probable flow demands, i.e. several appliances should be opened at once. Generally the pressure and flow performance tests are carried out by visual inspection, but it is possible to use a pressure gauge and flow measuring device to ensure the performance meets the required specification. A check should also be made at this stage for noise transmissions in the pipeline caused by rapid closure of valves.

All dhw thermostats will need to be inspected for correct operation and adjusted to maintain the water temperature as required, not exceeding 60°C.

(5) **Final System Check**s After any problems have been resolved, cistern lids should be secured, insulation material applied and labels fixed to valves for identification purposes, as necessary. The job should be left clean and tidy.

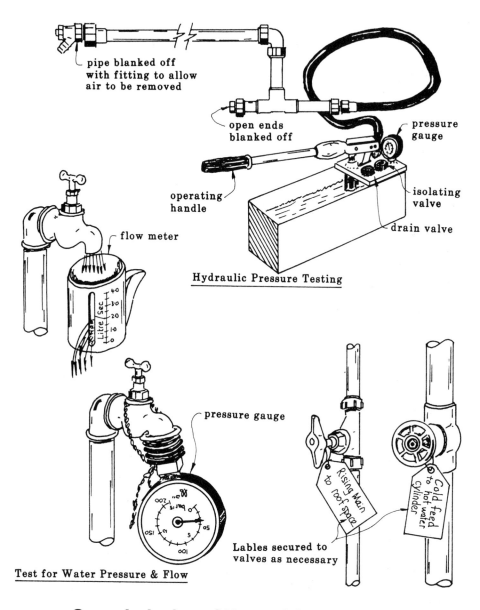

pipe blanked off with fitting to allow air to be removed

open ends blanked off

pressure gauge

operating handle

isolating valve

flow meter

drain valve

Hydraulic Pressure Testing

pressure gauge

Rising Main to roof space

Cold feed to hot water cylinder

Lables secured to valves as necessary

Test for Water Pressure & Flow

Commissioning of Hot and Cold Supplies

Maintenance and Servicing Schedule

Relevant British Standard
BS 6700

No system can be guaranteed for ever but its life expectancy can be greatly improved by identifying faults before they have a chance of causing any inconvenience. Planned preventive maintenance, regularly carried out, will not only help to ensure that the system performs as it was intended but may also prevent costly damage to equipment and buildings. A maintenance schedule is generally drawn up, and should be observed, giving guidance on what to look for when fulfilling the terms of a servicing contract. Shown is a typical schedule as used when inspecting a system of hot and cold water supplies.

Date of Inspection: _____
Inspected By: _____
Remarks: _____
_____
_____

Inspection carried out at:

Component	Remarks	Inspected	Notes
Meters	* Read meter and check water consumption for early signs of wastage * Confirm in correct working order		
Meter and stop valve chamber	* Ensure ease of opening to access covers * Clean out as necessary		
Earth bonding	* Check for alterations and suitable earth bonding maintained		
Water analysis	* '6 month' chemical and bacteriological analysis of drinking water systems where bulk storage exceeds 1000 litres		
Inspection covers and ducting	* Ensure ease of opening to covers and clean out service ducts as necessary * Check for signs of leakage, from pipework and surrounding ground or surface water * Check for the accumulation of gas		
In-line control valves	* Operate and confirm easy and effective operation * Labels clearly identify their purpose and are securely attached * Emergency valve keys readily available		
Terminal valves	* Check for suitable operation and effective closing * Remove scale build-up and clean spray-heads of shower mixers, etc. * Check timing delay of self-closing taps * Adjust water levels to float-operated valves		

Component	Remarks	Inspected	Notes
Pipework	* Check for suitable pressure and flow at outlets * Check supports and inspect for loose fittings * Check provision is maintained for expansion and contraction * Check for soundness of pipework * Inspect for signs of corrosion * Inspect insulation material for soundness * Inspect fire stopping to ensure that it is maintained		
Storage cisterns	* Confirm the cleanliness of vessels * Look for signs of leakage * Check for stagnant water (e.g. dust on surface of water) * Check condition of cistern supports * Confirm operation of overflow * Ensure lid and insulation is sound		
Pumps	* Check operation of any pump(s) fitted and ensure noise levels are minimal		
Pressure and temperature relief valves	* Open test lever to confirm valve not stuck down * Check discharge pipe not blocked		
Pressure-reducing valves	* Check pressures down-stream of valve		
Pressure vessels	* Inspect for corrosion and leakage * Drain vessels of water and measure gas pressure; adjust if necessary		
Filters	* Remove gauze/mesh trap and clean out		
Electrical components	* Check operation of all controls to include thermostatic devices * Check suitability of wiring to IEE standards		

Part 3
Central Heating

Domestic Central Heating

Relevant British Standard
BS 5449

The term *central heating* refers to a method of heating a building from a central heat source; it is this which distinguishes it from various forms of localised heating such as gas/coal fires or electric storage heating located within a room. The plumber is only concerned with wet systems of central heating; however, the warm air system is very often connected to primary flow and return pipework to a hot storage vessel. A design of central heating can also be found which uses such materials as electric cable embedded in the walls, designed to warm the structure; the concept is the same as that of radiant systems heating (see page 140).

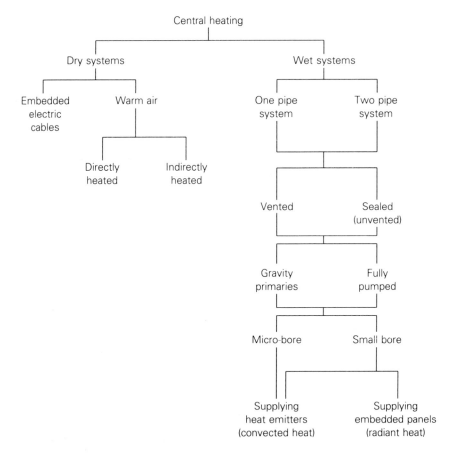

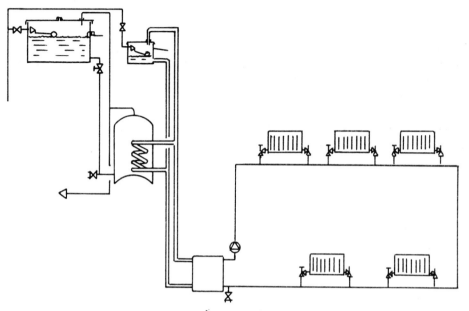

Small Bore 'One Pipe' Vented Heating System
with Gravity Primaries to the D.H.W.

Fully Pumped 'Two Pipe' Sealed Heating System
with Micro-bore to first floor Radiators

Domestic Central Heating

Wet Central Heating Systems

Relevant British Standard
BS 5449

Originally systems were designed to circulate the hot water by convection currents but because these systems had a very low circulating pressure larger pipe sizes were required. More modern systems use a pump to speed up this circulation. Because of this faster velocity it became possible to use smaller pipe sizes which led to smaller volumes of water in the system and thus a quicker heating up time. Installing a pump also enables freedom in system design; for example, the water can be pumped to circulate below the level of the boiler as convection currents are not relied upon. The water which is heated up in the boiler is often also used to circulate, either by pump or by convection currents, to an indirect hot water storage vessel, giving a domestic hot water supply.

One pipe system This system reduces installation costs, less pipework being required; it does, however, have several disadvantages when compared to the more expensive two pipe systems. *Firstly*, the first heat emitter in the system passes out its cooler water back into the main flow pipe; this results in the heat emitters at the end of the heating circuit being cooler than those at the beginning; therefore careful balancing of the system is essential. *Secondly*, because of its design, the pump only forces water around the main flow pipe and not through the individual radiators, these being heated only by convection currents; for this reason the heat emitter used must offer only a very little resistance to the natural upward flow of hot water.

Two pipe system With this system the water is not only pumped around the circuit but also through the radiators, giving them a much faster heating up period. Balancing out the heat to each radiator proves to be reasonably simple. It is not uncommon for the first radiator in the system to have its lock-shield valve just fractionally opened when balancing; this is due to the minimal frictional resistance to the flow through this heat emitter.

Two pipe reversed return system (three pipe system) This is a special design of the two pipe system in which the length of each heating circuit to each heat emitter in the system is about the same. When the cooler water leaves the first heat emitter in the system it does not simply join the return pipe and travel back to the boiler, as in the two pipe system; instead it travels to the furthest point in the system and upon receiving the return water from the last heat emitter runs back to the boiler return connection. This ensures that frictional resistance to the water flow is the same to each radiator; therefore, although the three pipe system can prove a little more complicated and expensive to install, balancing of the system proves to be a simple task.

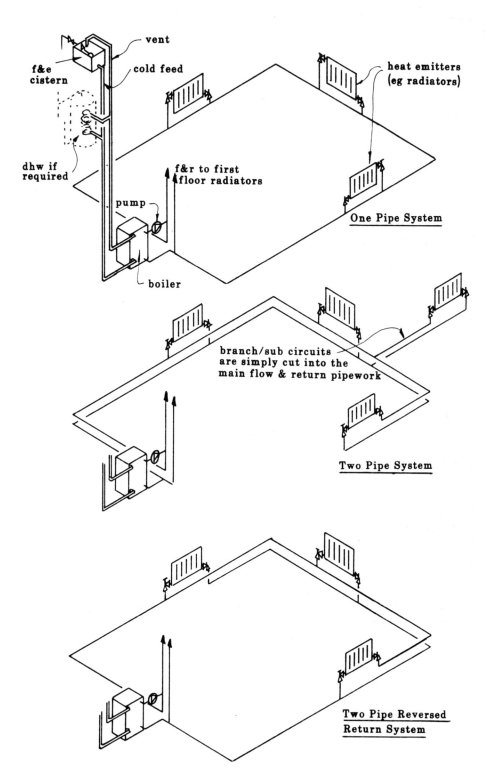

vent

f&e cistern

cold feed

dhw if required

heat emitters (eg radiators)

f&r to first floor radiators

pump

One Pipe System

boiler

branch/sub circuits are simply cut into the main flow & return pipework

Two Pipe System

Two Pipe Reversed Return System

Wet Central Heating Systems

Central Heating Components 1

Relevant British Standards
BS 2767, BS 3528 and BS 7556

Heat emitters

There are two main types of heat emitter: the radiator and the convector. Radiators, despite their name, give off very little radiant heat; they simply expose a large hot surface to the room, convection currents are set up, and the room is heated.

Radiators will be found in several designs (see figure). One of these, the convector radiator, has fins or plates welded to its back side which improve its ability to warm the air, exposing a greater surface to the air flow. The plates are heated by conduction from the hot radiator surface.

Convectors have a relatively small finned heating surface through which hot water passes. A series of closely attached fins warm up by conduction through which the air passes and is warmed. Two basic designs of convector will be found: those relying on the setting up of natural convection currents and those which use a fan to assist air circulation.

Radiator valves

These are valves fitted to the heat emitters; generally one valve is fitted to each end. The valves are identical except that one is fitted with a lock-shield head; this prevents unauthorised people from tampering with the regulated flow of water. The lock-shield valve is used only when balancing out the heating system, ensuring an equal distribution of hot water; if for any reason a radiator has to be removed from the wall, the valve can be shut down. The other radiator valve has a wheel head which can be used to turn the heater on and off.

Sometimes thermostatic radiator valves (trvs) are used to open and close the hot supply automatically as the room requires heat. The valve is fitted with a built-in heat sensor, or the sensor can be fitted in a better position away from the valve; remote sensors prove useful if, for example, the radiator valve is often covered with a curtain. As the sensor heats up a volatile liquid expands and is forced into the bellows chamber; this causes it to expand and exert a pressure on a pin, closing the valve.

When installing thermostatic radiator valves into a system it is essential that not all the heat emitters are fitted with a means of thermostatic control, unless a bypass is incorporated: should they all close at once no heat would be able to escape from the boiler and the pump could be pumping against this closed supply, possibly causing unnecessary damage. Should a radiator need to be removed at any time it is generally necessary to remove the temperature sensing head and secure the pressure pin down with a special manual locking nut; otherwise, should the temperature drop in the room, the valve may open, resulting in the discharge of water onto the floor.

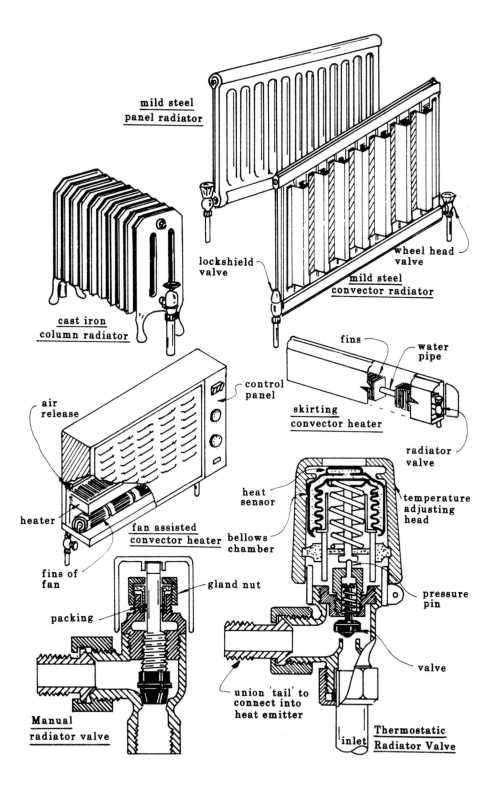

mild steel
panel radiator

cast iron
column radiator

lockshield
valve

wheel head
valve

mild steel
convector radiator

fins

water
pipe

skirting
convector heater

radiator
valve

air
release

control
panel

heater

fan assisted
convector heater

fins of
fan

heat
sensor

bellows
chamber

temperature
adjusting
head

gland nut

packing

union 'tail' to
connect into
heat emitter

pressure
pin

valve

Manual
radiator valve

inlet

Thermostatic
Radiator Valve

Central Heating Components 1

Central Heating Components 2

Feed and expansion cistern (f & e) A small cistern located at the highest point of a vented system of central heating. The cistern is designed to act initially as the fill-up point for the system, but its prime function is to allow the water in the system to expand. For this reason on initial fill-up the water level must be adjusted low down, just above the cold feed outlet into the system. The cistern must be placed in such a location that it will not be affected by the position and head of the circulating pump. A minimum dimension of the maximum head developed by the pump, divided by three, needs to be maintained between the water level and the pump to prevent undue water movement in the f & e cistern (see figure). When a c.h. system is connected to an indirect dhw storage cylinder the f & e cistern should ideally be located just below the water level of the dhw storage cistern; thus, if a split in the heat exchanger occurs and the c.h. and dhw systems mix, the fault will be identified by water discharging out from the f & e cistern overflow.

Circulating pumps These devices, sometimes called accelerators, are fitted to the pipework to assist water circulation. Basically pumps utilise a circular veined wheel which draws water in through its centre and throws it out by centrifugal force. The water velocity should not be too fast as noise will be generated; generally the velocity should not exceed 1 m/s for small bore systems and 1.4 m/s for micro-bore systems. The duty load of a pump should overcome the resistance of the index circuit (the circuit offering the greatest frictional resistance to flow). The location of a pump should, if possible, be such that it gives a positive pressure within the circuit, thus ensuring no air is drawn into the system via micro leaks (e.g. air being sucked in through radiator gland nuts).

Automatic air release valves A specially designed valve which enables air to escape from the system by allowing a float to rise and fall with the water level in the system. If water is present the float will be held up, forcing a washer against the outlet seating. When these valves are fitted to heating circuits they must only be installed in 'positive' flow pipework; otherwise, if installed on the negative (sucking) side of a pump, air will be drawn into the system when the pump is running.

Anti-gravity valve A valve fitted vertically in the pipeline, designed to overcome unwanted gravity circulation in central heating pipes. During the summer months when the boiler is used to heat the dhw the radiators will sometimes get hot due to gravity circulation. The valve will only open when pressure is created by a pump. The pressure exerted by convection currents is insufficient to cause the valve to lift. Fully pumped systems incorporating motorised valves do not require these valves to be fitted.

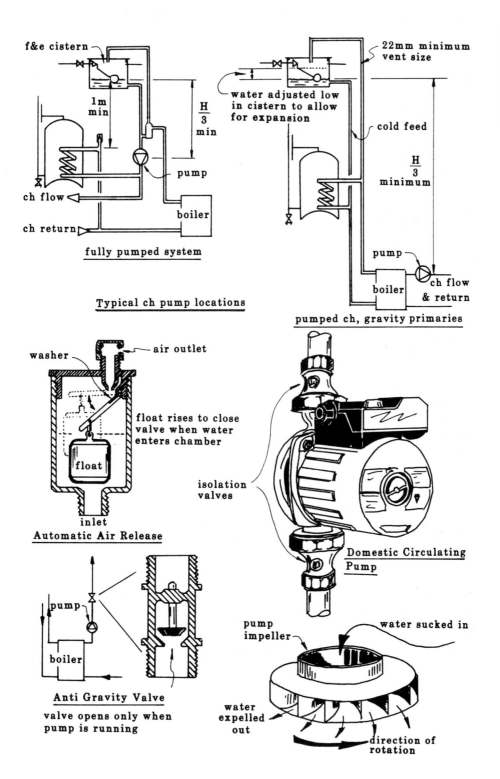

f&e cistern

1m min

H/3 min

pump

ch flow

ch return

boiler

fully pumped system

Typical ch pump locations

22mm minimum vent size

water adjusted low in cistern to allow for expansion

cold feed

H/3 minimum

pump

boiler

ch flow & return

pumped ch, gravity primaries

washer

air outlet

float rises to close valve when water enters chamber

float

inlet

Automatic Air Release

isolation valves

Domestic Circulating Pump

pump

boiler

Anti Gravity Valve

valve opens only when pump is running

pump impeller

water sucked in

water expelled out

direction of rotation

Central Heating Components 2

Heating Controls

Motorised valves

A valve fitted on top with an electrically operated motor which opens or closes the pipeline automatically. The power switching on the supply to operate the valve is regulated by either a thermostat or a time clock. There are two basic types of motorised valve, these being a *zone valve* and a *diverter valve*. The zone valve simply opens and closes the waterway and is fitted in a straight run of pipe, whereas a diverter valve is fitted at a 'tee' connection and sends the flow of water either one way or the other; this valve can be wired to give priority to either the domestic hot water or the heating circuit. Many diverter valves are designed to have a midway position, allowing water to flow in both directions at the same time. Systems using one zone valve are sometimes referred to as 'C' plan systems; when two zone valves are used the system is called the 'S' plan and where a diverter valve is incorporated it is called the 'Y' plan. See page 238 Central Heating Wiring Systems.

Programmer

This is a device consisting of a time clock which automatically switches on and off the boiler and pump and other controls to enable the user to override the time clock settings. The programmer also allows users to control the system so that it only heats the water they require; for example, to heat up only such domestic hot water as may be required during the summer months.

Room thermostat

This is a thermostat designed specifically to control the temperature within a building. When the desired room temperature is achieved the thermostat breaks a switch contact turning off a pump, boiler, or some other control and thus preventing the flow of heat to the room. A room thermostat should not be located where it would be affected by extreme temperatures; for example, in cold draughts or near any heat source. A good location would be in a living room or lounge at a height of about 1.6 m. It is best not to position a room thermostat in a hall, because the temperature is not critical here, and halls are often subject to draughts.

Cylinder thermostat

This thermostat is designed to control the temperature of the dhw. It is clamped to the outside wall of a hot storage cylinder and makes or breaks the electrical circuit, usually to a motorised valve, allowing or preventing the circulatory flow of primary water from the boiler.

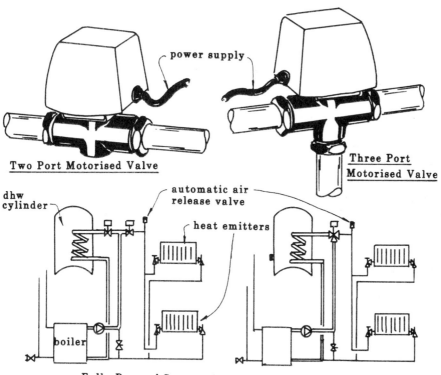

power supply

Two Port Motorised Valve

Three Port Motorised Valve

dhw cylinder

automatic air release valve

heat emitters

boiler

Fully Pumped Systems Incorporating Motorised Valves

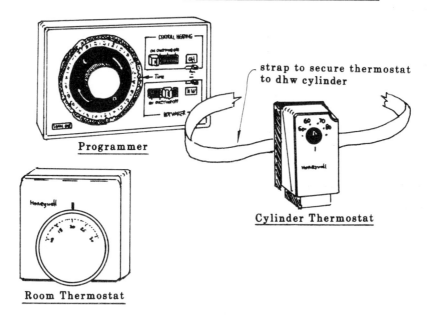

Programmer

strap to secure thermostat to dhw cylinder

Cylinder Thermostat

Room Thermostat

Heating Controls

Fully Pumped System

This is a system which operates fully under the influence of a pump. It does not rely at all upon convection currents being set up to circulate hot water to the hot storage vessel; thus it can be designed with its boiler anywhere, be it above or below the hot storage vessel. The motorised valves may be fitted either on the flow or return pipework and are generally wired up to the cylinder and room thermostats and made to automatically close when the required temperature within the cylinder or room has been reached.

When designing a vented fully pumped system supplied via an f & e cistern, care must be taken to locate the circulating pump in such a position as to ensure no positive or negative pressure at the vent pipe which will lead to pumping water over or sucking in air from this open-ended pipe. The position at which the cold feed enters the system is regarded as the neutral point and from this point to the pump it will be under a sucking, or negative influence. From the pump back to the cold feed inlet a pushing or positive force will occur. The ideally designed system should have a positive flow, pressurising the system. If the vent pipe connection is within 150 mm of the cold feed connection the vent will also be located at the neutral point and as a result no problems should occur regarding pumping water over or sucking in air.

An **air separator** is sometimes used. This allows cold feed and vent connections to be closely grouped; it also causes a turbulence of water flow in the pipe run which allows the formation of air bubbles which can simply rise up and out of the system, eliminating unnecessary corrosion problems.

Reversed circulation

This sometimes occurs in fully pumped systems in which some radiators get hot when not required (see figure). This is the result of a pressure difference between the two tee fittings at A and B when valve 'X' is closed. A water flow is set up, by-passing the closed motorised valve. This situation is simply avoided if the flow to the radiator circuits is split after the flow to the dhw cylinder and rejoins together before the dhw return connection.

A shorter heat recovery time can be achieved with fully pumped systems and it can be shortened even further by taking the primary flow into the lowest connection of the heat exchanger coil of the dhw cylinder. This allows heat transference by conduction and convection to be increased, because the hot surface is directly touching the coldest water in the cylinder.

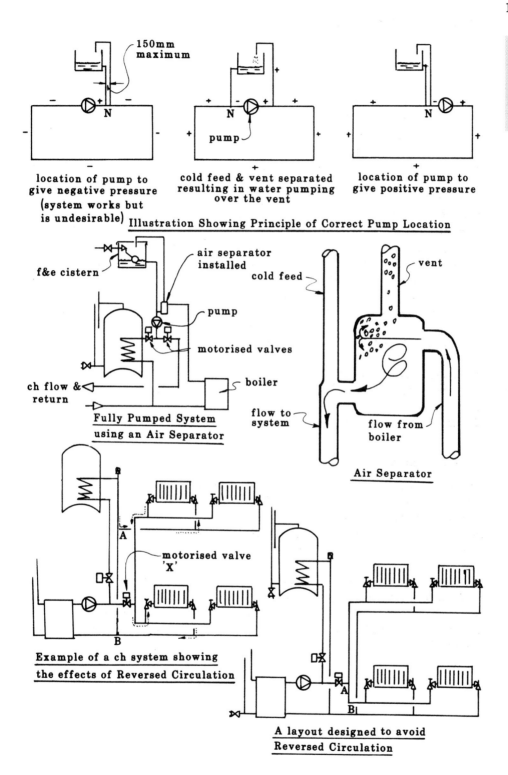

150mm maximum

location of pump to give negative pressure (system works but is undesirable)

cold feed & vent separated resulting in water pumping over the vent

location of pump to give positive pressure

Illustration Showing Principle of Correct Pump Location

f&e cistern

air separator installed

cold feed

pump

motorised valves

boiler

ch flow & return

Fully Pumped System using an Air Separator

vent

flow to system

flow from boiler

Air Separator

motorised valve 'X'

Example of a ch system showing the effects of Reversed Circulation

A layout designed to avoid Reversed Circulation

Fully Pumped System

Sealed Systems

Relevant British Standards
BS 4814 and BS 5449

These are hot water heating systems in which the water supplied to the system is fed from the supply main via a temporary supply pipe; connected to the circuit is a double check-valve assembly or some other means of preventing back-siphonage. Once the system is filled with water the inlet supply is shut off; thus all the water is entrapped in the circulating pipework. Note that the temporary fill connection should be removed after filling. When the boiler is fired the water heats up and expands and because this water (which is expanding) cannot be taken up in a feed and expansion cistern it generates a pressure on the internal pipework. Eventually this pressure acts upon a diaphragm and compresses the air and nitrogen gas located in a sealed expansion vessel, thus taking up the expansion of the water.

To allow for water replacement, which may be necessary because of leakage or venting, etc., either the temporary hose has to be replaced or the system is charged up to a pressure slightly in excess of the expansion vessel pressure; thus the make-up water will be taken up in the vessel itself, but care must be taken to ensure sufficient capacity in the expansion vessel to allow for the expanding water. Alternatively, a make-up cistern can be located above the highest point of the system which in turn can be filled either manually or connected to the supply main. The expansion vessel should be connected to the system on the inlet side of any pump, thus preventing the exertion of positive pressures on the diaphragm; also, the vessel should be located on the cooler return pipe to give a longer life.

A pressure gauge is installed to indicate the fill and system pressure. This should ideally be located close to the expansion vessel and fill connection. At no time must the temperature of these systems be allowed to exceed 100°C; therefore a high temperature cut-out device must be included, in addition to the normal boiler thermo-stat. To prevent excessive pressures building up within the system a pressure relief valve is fitted to open at 2–3 bar pressure. Any discharging water should be safely conveyed to a suitable drain point, via a tundish (a funnel-shaped pipe which must maintain an effective air gap).

It must be noted that because a pressure is created in this type of system higher water temperatures can be achieved owing to the fact that water boils at a higher temperature under pressure. Therefore convector heaters or heating panels are often chosen in preference to panel radiators to prevent anyone from being scalded. The advantages of sealed systems over more conventional systems include:

☐ Less pipework is necessary on installation;
☐ The pumping of water over vent pipes or drawing air into the system as in fully pumped systems is eliminated;
☐ Higher water temperatures can be achieved;
☐ The boiler can be positioned anywhere, even in the roof space, as no header cistern is required.

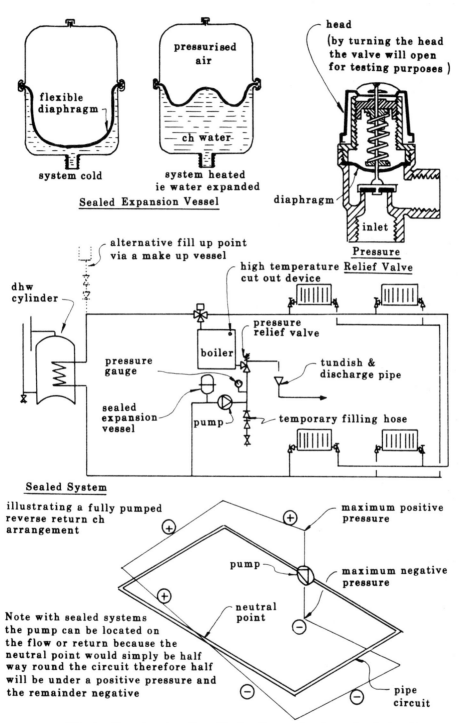

flexible diaphragm

system cold

pressurised air

ch water

system heated ie water expanded

Sealed Expansion Vessel

head
(by turning the head the valve will open for testing purposes)

diaphragm

inlet

Pressure Relief Valve

alternative fill up point via a make up vessel

dhw cylinder

high temperature cut out device

pressure relief valve

boiler

tundish & discharge pipe

pressure gauge

sealed expansion vessel

pump

temporary filling hose

Sealed System

illustrating a fully pumped reverse return ch arrangement

maximum positive pressure

pump

maximum negative pressure

neutral point

Note with sealed systems the pump can be located on the flow or return because the neutral point would simply be half way round the circuit therefore half will be under a positive pressure and the remainder negative

pipe circuit

Illustration showing the effects of pumping in a closed circuit

Sealed Systems

Boilers

Relevant British Standards
BS 5449 and BS 6798

Solid fuel boilers These include those which burn such materials as wood, straw and coal. The design can vary tremendously from the 'pot' burner, in which the heat exchanger (area containing the water) surrounds the combustion chamber, to a design which burns smokeless fuels such as anthracite pellets (hard coal) which are automatically fed into the burner, via a hopper, and the heat is given off as it rises through a series of waterways.

Gas boilers The heat exchanger of a gas boiler is a close network of waterways through which the hot gases rise. No great fear is given to the sooting-up of this heat exchanger because of the cleanliness of the fuel and its combustion process.

Oil boilers These can be found in several designs although only those using a pressure jet burner are produced today. Oil burners can produce excessive carbon deposits (soot); therefore the heat exchanger needs to be such that it is easily cleaned out. It consists of a chamber surrounded by the waterway, the heat being directed on to the walls by a series of baffles.

Electric storage boilers These are a new concept which use cheap rate night-time electricity to heat an element which warms up a series of refractory blocks. During the day when the heat is required, a fan blows air around a closed circuit which warms and in turn blows on to a water-filled heat exchanger.

Boiler noises and design considerations Noises created in the boiler are often due to the formation of scale, especially in the flow pipe; air is entrapped by the scale and a kind of boiling noise (often called kettling) ensues, caused by steam forming and condensing. The scale can generally be removed by treating the system with a descaling solution. With solid fuel boilers and low-water-content boilers it may be necessary to have a heat leak from the boiler to allow the dissipation of residual heat. A heat leak may be any heat emitter, or sometimes the dhw cylinder is used, providing the circuit cannot be closed to allow the heat to circulate and thus escape from the boiler.

In the specification of any boiler one needs to consider the heat *input* and heat *output*. The heat input is the result of the fuel being consumed whereas the heat output is the energy produced and available for use. For example, a 75% efficient gas boiler which has a 17 kW input will only provide 12.75 kW output to the system. When commissioning the system, after the initial warm up, the water should never be allowed to flow back to the boiler below 60°C, as condensation will form in the heat exchanger, resulting from the combustion of the fuel, and will cause corrosion problems, shortening the life of the boiler appreciably.

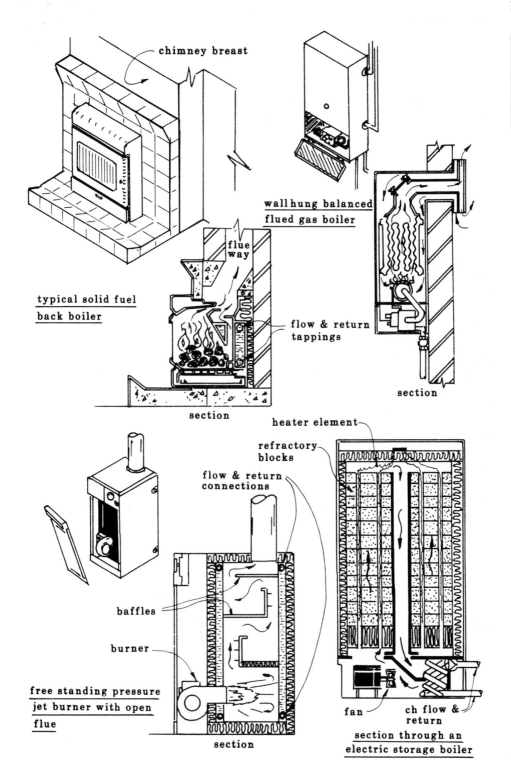

chimney breast

wall hung balanced
flued gas boiler

flue
way

typical solid fuel
back boiler

flow & return
tappings

section

section

heater element

refractory
blocks

flow & return
connections

baffles

burner

free standing pressure
jet burner with open
flue

fan

ch flow &
return

section through an
electric storage boiler

section

Boilers

Combination Boiler (Combi)

This is a specially designed boiler which is used to heat up the domestic hot water instantly, as and when required, and also to serve a system of hot water central heating. The combination boiler reduces installation costs because no feed and storage vessels are required for the supply of water; also, by omitting the storage of domestic hot water, this boiler saves money which might have been spent heating the water unnecessarily.

Combination boilers are only suitable for low-energy homes in which the demand for hot water is limited. There is a limit both to the volume of water and the speed with which it can be heated up, and also, during this time of dhw demand, the c.h. is turned off. When considering whether or not to install a combination boiler one needs to ensure that the flow rate of water is sufficient to supply the volume of water needed and possibly, in the case of mains supply systems, to allow for all hot and cold draw-off points. Therefore, if several taps are opened at once it may lead to some appliances being starved of water.

There are many variations of combi, all working on different design concepts. One such system operates as follows:

(1) Should the central heating system call for heat the pump is energised. This starts the water flowing. As the water passes through a venturi a pressure differential occurs in the deficiency valve causing the gas valve to open.
(2) Gas flows through the main burner and is ignited by the pilot flame.
(3) The water is rapidly heated in the low-water-content heat exchanger and can only circulate around the boiler, through the dhw heat exchanger. The expansion of the water is taken up in the sealed expansion vessel.
(4) As and when the temperature of the water reaches 55–60°C the thermostatic element expands, causing the hot water system valve to close and the heating system valve to open; this allows water to flow around the heating circuit.
(5) The closing of the hot water system valve also causes a rod to rise and activate a micro-switch.This notifies the boiler control box that higher temperatures can be achieved which are manually determined by the setting of the flow temperature selector, on the control panel, and range from 60–90°C.

Should domestic water be required the following operation takes place:

(1) When a hot draw-off point is opened water flows through the differential pressure valve; this causes the diaphragm to lift and activate a micro-switch, energising the pump.
(2) The main burner ignites and the boiler functions, as in (1), (2) and (3) above.
(3) As the cold water passes over the thermostatic element it keeps it cool, ensuring that it does not expand, and causes the heating system valve to open. Therefore the central heating hot water only circulates through the dhw heat exchanger and around the boiler.
(4) As the cold water passes over the dhw heat exchanger it is rapidly heated before being discharged through the hot tap.

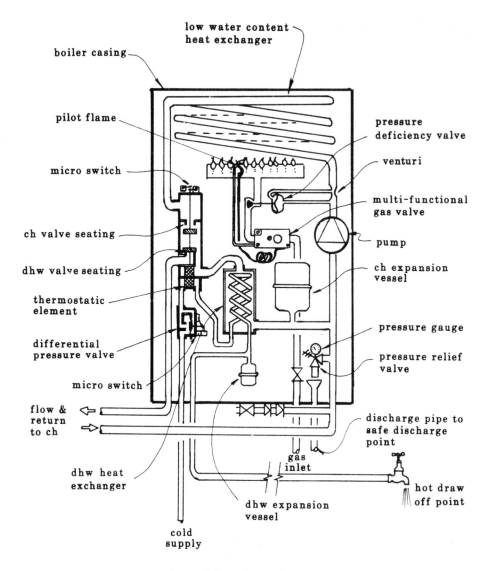

low water content
heat exchanger

boiler casing

pilot flame

micro switch

ch valve seating

dhw valve seating

thermostatic
element

differential
pressure valve

micro switch

flow &
return
to ch

dhw heat
exchanger

cold
supply

pressure
deficiency valve

venturi

multi-functional
gas valve

pump

ch expansion
vessel

pressure gauge

pressure relief
valve

discharge pipe to
safe discharge
point

gas
inlet

hot draw
off point

dhw expansion
vessel

Combination Boiler

Condensing Boiler

This is a design of boiler which can have an increased efficiency over the more traditional boiler. The efficiency of a typical non-condensing boiler is around 75%, whereas with condensing boilers a figure of over 87% can be expected. This increased efficiency is due to the extraction of heat from the otherwise wasted flue gases. Most boilers have a single combustion chamber enclosed by the waterways of the heat exchanger through which the hot gases can pass. These gases are eventually expelled through the flue, located at the top of the boiler, at a temperature of around 180°C.

Condensing boilers, on the other hand, are designed first to allow the heat to rise upwards through the primary heat exchanger; when at the top the gases are rerouted and diverted back down over a secondary heat exchanger. These can reduce the flue gas temperature to about 55°C. This reduction of temperature causes the water vapour (formed during the combustion process) to condense and, as the droplets of water form, fall by gravity to collect at the base of the flue manifold. The remaining gases are expelled to the outside environment through a fan-assisted balanced flue. The condensation produced within the appliance should be drained as necessary into the waste discharge pipework or externally into a purpose-made soakaway.

It is only possible for a condensing boiler to work to these very high efficiencies if the flow and return pipework is also kept below 55°C. These low f & r temperatures need to be maintained for the heat transference to occur from the flue to the water (i.e. heat transference goes from hotter to cooler materials).

Many people are installing condensing boilers in homes which are fitted with radiators and a primary flow and return to the dhw. Some of these people may be under the impression they are getting more for their money; unfortunately, as stated above, low flow and return connections are essential; therefore they are not making the vast savings they are led to expect. For a c.h. system to work with radiators and dhw primary circuits, flow temperatures need to be around 82°C, so in fact the installer has put in an expensive condensing boiler which gives only slightly improved efficiency over the more traditional boiler. The appliance basically only works in its condensing mode during initial heat-up.

To achieve a system which will function in its condensing mode possibly the installer needs to consider a suitable system of radiant heating. This will be identified and discussed later in the book.

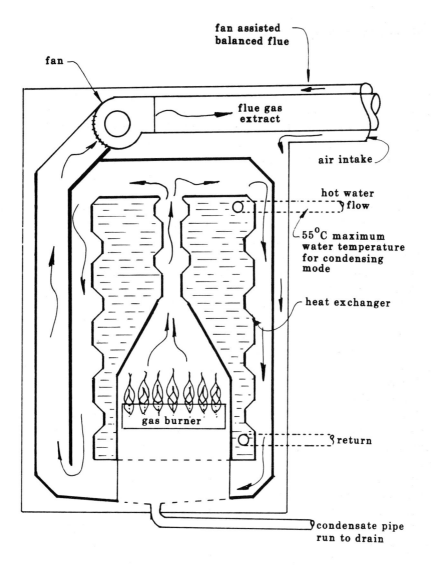

fan assisted
balanced flue

fan

flue gas
extract

air intake

hot water
flow

55°C maximum
water temperature
for condensing
mode

heat exchanger

gas burner

return

condensate pipe
run to drain

Illustration showing the principle of
operation to a Condensing Boiler

Condensing Boiler

Radiator and Boiler Sizing

Relevant British Standard
BS 5449

Correct heat emitter sizing can be achieved using a c.h. calculator such as the *Mears*, a computer program or mathematical calculation, as shown here. Firstly one must find the *heat lost from the room*, which occurs in either of the following two distinct ways:

(1) *Heat loss due to air change and natural ventilation.* This is found using the formula:

| Volume of room (m³) | × | Air change (per hour) | × | Temperature difference (°C) | × | Ventilation factor (0.33 W/m °C) | = | Heat loss (W) |

The following *air change rates* should be allowed for: two per hour in kitchens, bathrooms and dining areas and one and a half per hour for other rooms.

(2) *Heat loss through the building fabric.* This is found using the formula:

$$\text{Surface area (m}^2) \times \text{Temperature difference (°C)} \times \text{U value (W/m}^2 \text{ °C)} = \text{Heat loss (W)}$$

The temperature difference This is the difference between the internal and external environment. The external temperature is usually taken to be $-1°C$, although a colder outside temperature may be allowed for in exposed locations. The internal temperature is to the client's needs, usually based on the following: $21°C$ for living, dining, bath and bed sitting rooms and $16–18°C$ for kitchen, hall, WC, and bedrooms. Note that where the dwelling adjoins another (e.g. a semi-detached) one assumes a temperature difference of $6°C$.

The U value This is found from tables, such as those listed in the Chartered Institute of Building Services, or Institute of Plumbing Design Guides. It is possible to use the following approximate values without knowledge of the correct U value.

Approximate U value through building fabric

Construction	W/m²°C	Construction	W/m²°C
External solid wall	2.0	Ground floor, solid	0.45
External cavity wall	1.0	Ground floor, wood	0.62
External cavity wall (filled)	0.5	Intermediate floor, heat flow up	1.7
External timber wall	0.6	Intermediate floor, heat flow down	1.5
Internal wall	2.2	Flat roof	1.5
Single glazing	5.7	Pitched roof (100 mm insulation)	0.34
Double glazing	3.0	Pitched roof (no insulation)	2.2

Heat emitter sizing If one finds the total heat loss (in watts) from a room and installs a heater giving the same heat output, the temperature will be maintained. To allow a cold room to warm up requires the heater to be increased in size by a small percentage (usually 15%) although this is not applicable if the heating is on for 24 hours a day. With reference to the bungalow in the figure, find the heat emitter requirements for the lounge and the bedroom. The answer is on page 134.

<u>Notes:</u>

All dimensions in metres

solid external brick wall

solid floor

single glazed windows

double glazed doors

100mm insulation in roof space

3kW to be allowed for dhw

room heights 2.4m

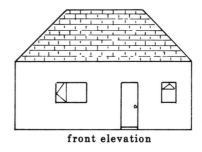

front elevation

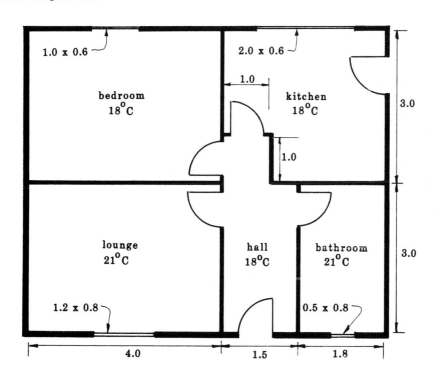

Radiator and Boiler Sizing

Heat emitter sizing (cont'd)

Heat requirements. Location: lounge

Fabric loss element	Area $L \times B = (m^2)$		Temperature difference (°C)		U value $(W/m^2{}°C)$	Heat loss (W)
Window	$1.2 \times 0.8 = 0.96$	×	22	×	5.7	120.38
External walls	$7 \times 2.4 = 16.8 - 0.96 = 15.84$	×	22	×	2.0	696.96
Internal walls	$7.0 \times 2.4 = 16.8$	×	3	×	2.2	110.88 +
Floor	$4.0 \times 3.0 = 12.0$	×	22	×	0.45	118.8
Roof	$4.0 \times 3.0 = 12.0$	×	22	×	0.34	89.76
						1136.78

Ventilation loss

volume × air change × temperature difference × factor   +

$3 \times 4 \times 2.4 \times$ 1.5   ×   22   × 0.33   =   313.63

1450.41

Plus 15% for intermittent heating =   217.56 +

1667.97

Heat requirements. Location: bedroom

Fabric loss element	Area $L \times B = (m^2)$		Temperature difference (°C)		U value $(W/m^2{}°C)$	Heat loss (W)
Window	$1.0 \times 0.6 = 0.6$	×	19	×	5.7	64.98
External walls	$7.0 \times 2.4 = 16.8 - 0.6 = 16.2$	×	19	×	2.0	615.6
Internal walls	no heat losses	×	n/a	×	n/a	0.0 +
Floor	$4.0 \times 3.0 = 12.0$	×	19	×	0.45	102.6
Roof	$4.0 \times 3.0 = 12.0$	×	19	×	0.34	77.52
						860.7 −
Fabric gain						
Internal walls	$4.0 \times 2.4 = 9.6$	×	3	×	2.2	63.36
						797.34

Ventilation loss

volume × air change × temperature difference × factor   +

$3 \times 4 \times 2.4 \times$ 1.5   ×   19   × 0.33 =   270.86

1068.20

Plus 15% for intermittent heating   =   160.23 +

1228.43

Note: There is a heat gain to the bedroom from the lounge.

Having found the heat required in watts, one simply refers to a manufacturer's radiator catalogue, as in the following example, to find the size of heat emitter required. From the schedule, the lounge will require a 1600 mm long × 590 mm high single convector, or a 960 mm × 590 mm double convector may be chosen. The bedroom will require a 1280 mm long × 590 mm single convector or a 800 mm × 590 mm double convector radiator.

135

3 Central Heating

Sample section from a radiator schedule

Convector radiators				Height 23 in (590 mm)		Tappings 4 × ½ in		
Single convectors				Length		Double convectors		
Order code	Heat emission		Price			Order code	Heat emission		Price
	Btu/h	W	£	in	mm		Btu/h	W	£
23 SC 12	1662	487	18.20	18.9	480	23 DC 12	2969	870	41.50
23 SC 16	2235	655	27.50	25.2	640	23 DC 16	4009	1175	55.50
23 SC 20	2805	822	33.70	31.5	800	23 DC 20	5047	1479	68.80
23 SC 24	3378	990	39.80	37.8	960	23 DC 24	6087	1484	82.00
23 SC 28	3951	1158	46.00	44.1	1120	23 DC 28	7128	2089	91.50
23 SC 32	4521	1325	52.40	50.4	1280	23 DC 32	8169	2394	102.20
23 SC 36	5094	1493	59.50	56.7	1440	23 DC 36	9209	2699	115.90
23 SC 40	5668	1661	65.80	63.0	1600	23 DC 40	10250	3004	125.60
23 SC 44	6237	1828	72.70	69.3	1760	23 DC 44	11291	3309	138.50
23 SC 48	6811	1996	79.50	75.6	1920	23 DC 48	12328	3613	151.70
23 SC 52	7384	2164	84.00	81.9	2080	23 DC 52	13369	3918	162.90
23 SC 56	7954	2331	90.30	88.2	2240	23 DC 56	14409	4223	167.80
23 SC 60	8527	2499	96.50	94.5	2400	23 DC 60	15450	4528	175.40

Boiler Sizing

The required boiler output is determined as follows:

(1) Total up the fabric requirements for all rooms
(2) Add half the sum total of ventilation heat losses*
(3) Add the domestic heat requirements (e.g. 1 kW for every 50 litres)
(4) Add a 20% margin for heat loss from pipes and initial warm up.

* Note: the total ventilation heat loss is not included because some air change will be the result of warm air passing from one room to another; thus it is, in effect, a heat gain.

Given the fabric and ventilation heat losses from the bungalow previously identified, and allowing for a 150 litre dhw cylinder, the boiler output would need to be 8½ kW, as the calculation here shows.

Boiler size for bungalow

Room	Vent loss (W)	Fabric loss (W)
Lounge	313.63	1136.78
Bedroom	270.86	797.34
Kitchen	294.94 +	794.40 +
Hall	124.15	130.33
Bathroom	188.18	709.32
	1191.76 ÷ 2	3568.17
half vent loss 595.88		→ 595.88 +
		4164.05
dhw requirement		→ 3000.00 +
		7164.05
20% margin		→ 1432.81 +
		8596.86 W
		(8½ kW)

Pipe and Pump Sizing

Relevant British Standard
BS 5449

The size of the pipe and pump required to serve the heat emitters in domestic situations is generally based on rule of thumb and general experience, which in most cases works sufficiently well. However, for the larger job or where efficiency is paramount the size of the pipe and pump may be calculated.

Shown opposite is a completed example of a small heating system and the calculation table. Note that the pipe size is indicated in the calculation table; the pump size is deduced from its results (see page 139). The stages and interpretation of the results may be explained as follows:

Column 1 This is the section of pipework which is being sized; note that the system is broken down into various sections.

Column 2 The total required heat emitter values are inserted here, taking account of all the heat losses and intermittent heating (see pages 132–135 on radiator sizing).

Column 3 An allowance of 10–25% of the heat emitter size is inserted to allow for heat loss from the pipes due to standards of insulation pipe runs, etc. I have allowed 20%. Example in section A 20% of $11.1 = (20 \div 100 \times 11.1) = 2.22$.

Column 4 This is the sum totals of columns 2 and 3; thus in section A $11.1 + 2.22 = 13.32$, this being the total heat requirement for the section.

Column 5 Generally a constant figure of 46 can be inserted here. The figure 46 has been arrived at by multiplying the specific heat of water (4.186 kJ/kg$^\circ$C) by the system design temperature drop. I have assumed 82°C flow and 71°C return; thus $82 - 71 = 11^\circ$C. Therefore, by calculation, $4.186 \times 11 = 46$.

Column 6 This is found by dividing column 4 by column 5; thus in section A $13.32 \div 46 = 0.29$, this being the amount of flow, in litres per second, required to maintain the required heat emissions.

Column 7 The actual pipe size required is found by referral to the table on page 138. The flow rate is given within the table for various diameters of pipe. When using the table ideally a pipe size should be chosen which allows a flow rate within the range 200–400 pascals/metre (Pa/m) for economic reasons. (Where the pressure is much less than 200 Pa/m the pipe is likely to be oversized, which increases the installation costs. Conversely, where the pressure is much greater than 400 Pa/m the pipe would be undersized; a larger pump will be required and running costs will increase.)

Example: In section A where a flow rate of 0.29 l/s is required a pipe size of 22 mm or 28 mm may have been chosen. Neither actually falls within the 200–400 Pa/m band and both are at equal distances from the pressure range. I have selected the 22 mm pipe size for economic reasons, this size of pipe being cheaper to purchase and easier to install.

Pipe sizing to c.h. system and indication of system working pressure

1	2		3	4		5	6	7	8		9	10
Section	Total heat emitter value		20% addition for heat loss from pipes	Total heat requirements		Constant figure	Flow rate	Pipe size	Pressure loss		Total effective length	Pressure required
	(W)		(W)	(W)			(l/s)	(mm)	(Pa/m)		(m)	(Pa)
A	11.1	+	2.22	13.32	÷	46	0.29	22	440	×	30	13200
B	5.5	+	1.1	6.6	÷	46	0.14	22	120	×	27	3240
C	2.0	+	0.4	2.4	÷	46	0.05	15	140	×	10	1400
D	3.5	+	0.7	4.2	÷	46	0.09	15	360	×	20	7200
E	1.8	+	0.36	2.16	÷	46	0.05	15	140	×	15	2100
F	5.6	+	1.12	6.72	÷	46	0.15	22	140	×	38	5320
G	2.6	+	0.52	3.12	÷	46	0.07	15	240	×	19	4560

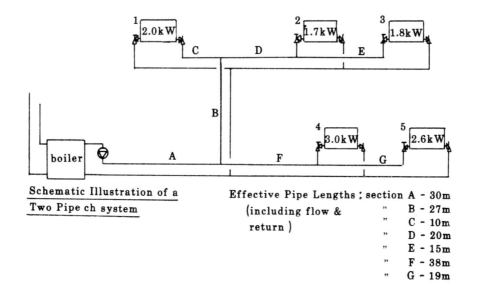

Schematic Illustration of a
Two Pipe ch system

Effective Pipe Lengths : section A - 30m
(including flow &   "   B - 27m
return)   "   C - 10m
                           "   D - 20m
                           "   E - 15m
                           "   F - 38m
                           "   G - 19m

Pipe and Pump Sizing

Flow of water at 75°C in copper pipes to Table X (BS 2871)

Pressure loss	Flow rate in litres/s					
	Pipe size					
Pa/m	12 mm	15 mm	22 mm	28 mm	35 mm	42 mm
90.00	0.021	0.040	0.118	0.239	0.430	0.725
92.50	0.022	0.041	0.120	0.242	0.437	0.735
95.00	0.022	0.042	0.122	0.246	0.444	0.748
97.50	0.022	0.042	0.124	0.250	0.450	0.759
100.00	0.023	0.043	0.125	0.253	0.457	0.769
120.00	0.025	0.047	0.139	0.281	0.506	0.852
140.00	0.028	0.052	0.152	0.306	0.551	0.928
160.00	0.030	0.056	0.164	0.330	0.594	1.00
180.00	0.032	0.060	0.175	0.352	0.635	1.07
200.00	0.034	0.064	0.186	0.374	0.673	1.13
220.00	0.036	0.067	0.196	0.394	0.710	1.19
240.00	0.038	0.071	0.206	0.414	0.745	1.25
260.00	0.039	0.074	0.215	0.433	0.779	1.31
280.00	0.041	0.077	0.224	0.451	0.812	1.37
300.00	0.043	0.080	0.233	0.469	0.844	1.42
320.00	0.044	0.083	0.242	0.486	0.874	1.47
340.00	0.046	0.086	0.250	0.503	0.904	1.52
360.00	0.048	0.089	0.258	0.519	0.913	1.57
380.00	0.049	0.092	0.266	0.535	0.962	1.62
400.00	0.050	0.094	0.274	0.551	0.990	1.66
420.00	0.052	0.097	0.282	0.566	1.02	1.71
440.00	0.053	0.099	0.289	0.581	1.04	1.75
460.00	0.055	0.102	0.297	0.595	1.07	1.80
480.00	0.056	0.104	0.304	0.610	1.10	1.84
500.00	0.057	0.107	0.311	0.624	1.12	1.88
520.00	0.059	0.109	0.318	0.637	1.15	1.92
540.00	0.060	0.112	0.324	0.651	1.17	1.96
560.00	0.061	0.114	0.331	0.664	1.19	2.00
580.00	0.062	0.116	0.338	0.677	1.22	2.04
600.00	0.064	0.119	0.344	0.690	1.24	2.08

(From CIBSE Guide Section C4, reproduced with permission of the Chartered Institution of Building Services Engineers.)

Column 8 Once the pipe size has been chosen the pressure loss is recorded (see the table above). Thus in section A where the flow rate was 0.29 and a 22 mm pipe was selected, the pressure loss reads 440 Pa/m.

Column 9 The total effective length is the run of flow and return pipework, including the actual pipe length and any additional length, due to fittings. See page 102 Pipe Sizing of Hot and Cold Pipework, for a worked example. In this example I have given the effective pipe runs and they are indicated on the figure.

Column 10 This is the actual pressure required for each section and is found by multiplying column 8 by column 9; thus in section A 440 × 30 = 13200 Pa.

Pipe sizing

In specifying the pump performance both the maximum flow rate (litres/s) and the maximum pressure (Pa) need to be stated.

☐ The maximum flow rate can clearly be seen in the completed example against the largest pipe size, to which 0.29 l/s is indicated.
☐ The maximum pressure is found by adding together the pressure from each section of pipework in a circuit. In our example there are three possible circuits to consider.

The circuit with the greatest pressure drop is known as the *index circuit* and in theory if you can circulate around this pipeline you can circulate around any part of the system. Calculation to find the index circuit:

1st circuit, to radiator No 1 (sections A, B and C)
2nd circuit, to radiator No 3 (sections A, B, D and E)
3rd circuit, to radiator No 5 (sections A, F and G)

1st circuit			*2nd circuit*			*3rd circuit*		
section A:	13200		section A:	13200		section A:	13200	
section B:	3240	+	section B:	2340		section F:	5320	+
section C:	1400		section D:	7200	+	section G:	4560	
	17840	Pa	section E:	2100			23080	Pa
				24840	Pa			

Thus we can conclude by identifying the index circuit as that from the boiler to radiator No 3, requiring a pressure of 25740 Pa (or 25.7 kPa). Note that the index circuit is not always the circuit with the greatest actual pipe length; this is dependent upon the amount of fittings used.

The final pump selection can now be made; our example requires a pump/circulator with a maximum pressure of 24.8 kPa and a maximum flow of 0.29l/s. By referral to the performance curve of a pump (given in the manufacturer's data), we would need a pump which meets our requirements, the pressure and flow falling within the curve itself. (Note: 1 kPa = 1 N/m².)

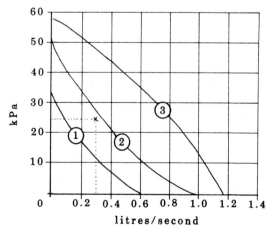

By referral to the performance curve the pump chosen may be selected on the number two setting

Pump Performance Curve

Radiant Heating

This is a system of heating designed to raise the temperature of a room space by the admission of infrared energy, which is basically thermal radiation. Radiant heat passes directly through air and will only heat the more solid surfaces upon which it falls. Radiant heat panels are usually mounted in the floor, walls or ceiling of a room. The panels are heated electrically or by circulating hot water or hot air through them. Unlike other forms of central heating, the effectiveness of radiant heating does not rely on the efficient circulation of air due to direct contact with the heat source.

For an equal state of comfort in a room, systems relying mainly on convection currents (such as a radiator type system) must provide a higher air temperature within the room because the cold surfaces of walls and windows, etc., remove heat from the human body that can only be replaced by the surrounding/ambient air. Radiant heat on the other hand warms up the floors, walls, windows, etc., and thus reduces the heat lost from the human body. Therefore, lower air temperatures within the room are maintained; this also provides a greater feeling of freshness, and with the reduction of convection currents in the room cold draughts and dust problems are reduced to a minimum.

Radiant heating can save fuel, unless the heating of the room is intermittent, as in, for example, a building which lowers its air temperature at night and requires it to be rapidly raised in the morning.

The design of radiant heating systems is quite straightforward, requiring only a coil of embedded pipe running through the surface of a floor, wall or ceiling; the only important requirement is that the surfaces of the radiant heaters should not be metallic. Fitted behind the heater should be a means of thermal insulation to give heat emission only into the room. With this system one major disadvantage is the danger of a pipe leakage which can prove difficult to find and expensive to repair.

In the following table is given the recommended surface temperatures of walls, etc., fitted with embedded pipe panels. Notice how panels located in the ceiling can be used to give off a higher radiant heat emission, and therefore give a quicker room heating period and are generally chosen. The water temperature flowing through the pipes should be between 40–55°C. To achieve the higher surface temperatures used in the ceiling, the heating pipes are spaced closer together.

Panel location	Surface temperature (°C)	Heat given off (%)
Ceiling	40	65
Floors/walls	24	50

For a radiant system to operate efficiently and effectively the system needs to be *on* constantly, thus preventing the building fabric from cooling. Because the temperature flowing through the pipework is greatly reduced, compared with the more traditional system using radiators with flow temperatures around 82°C, these systems can be designed with a condensing boiler which will increase the efficiency of the system even more.

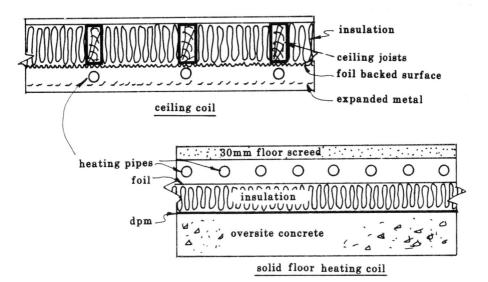

insulation

ceiling joists

foil backed surface

expanded metal

ceiling coil

heating pipes

foil

dpm

30mm floor screed

insulation

oversite concrete

solid floor heating coil

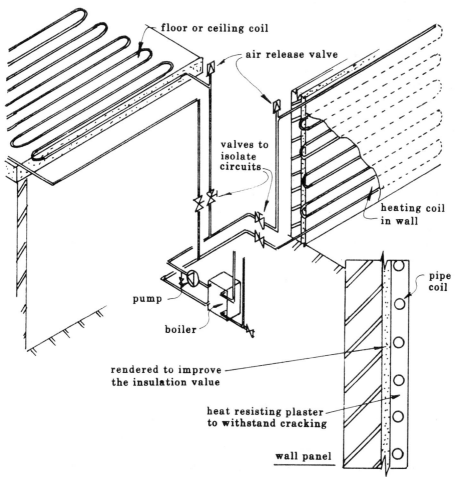

floor or ceiling coil

air release valve

valves to
isolate
circuits

heating coil
in wall

pump

boiler

pipe
coil

rendered to improve
the insulation value

heat resisting plaster
to withstand cracking

wall panel

Radiant Heating

Design Concepts in Central Heating

Relevant British Standard
BS 5449

Pump overrun A situation which allows the c.h. pump to continue running for a short period after the boiler has shut down. It is designed to remove any residual heat left in the boiler (e.g. low water content boilers) which would otherwise boil and may cause a noise or the ejection of water from an open vent. The boiler manufacturer will state in the installation instructions whether pump overrun is required.

Central heating by-pass Several applications may require the inclusion of a by-pass, for such purposes as a route of water flow in pump overrun, as described above. Other reasons may include:

☐ to assist in giving the 11°C design temperature drop across the system when balancing (see page 146 on commissioning), thus allowing hot water to flow directly back into the return close to the boiler. *Or*
☐ where thermostatic valves are fitted to all heat emitters and the dhw cylinder the by-pass gives a flow route when all valves are shut down.

The size of the by-pass should be 15 mm in diameter for boilers up to 18 kW and 22 mm for boilers exceeding 18 kW. A lock-shield headed valve will need to be fitted on the by-pass to regulate the flow of water.

External temperature control To save on fuel consumption sometimes a modulating 3-port valve is fitted to circulate a proportion of the c.h. return water to the flow (thus by-passing the boiler). Heating system design is usually based on a $-1°C$ outside temperature; therefore, for most of the time the system is oversized. This control monitors the outside environment and compensates for changes in temperature, and the flow temperature is automatically adjusted to suit. The heat emitters are sized for $82°C$ flow and $71°C$ return, and therefore by reducing this temperature the heat emission will be reduced as necessary. The capital installation cost of this control is high, and therefore rarely used in domestic situations.

Frost protection In general, most domestic c.h. systems have thermal insulation material applied to exposed pipework and f & e cisterns only as a minimum requirement. However, it is possible to employ one of the following methods to improve the protection of the system:

Set back control: Possibly the surest way to ensure that freezing does not occur. The system is allowed to operate 24 hours a day, with a reduction of $5°C$ in temperature at night. This system can be somewhat high on fuel consumption.
Frost thermostat: A thermostat located at the coldest point of the system which is wired up to override any other control and as a result will bring on the heating system. The thermostat is usually set to bring on the system at $2–3°C$. Sometimes a room thermostat is used continuously (24 h a day) and adjusted to $8°C$ at night; therefore, generally the heating will be off but during a cold spell the system will switch on. This set-up is based on an assumption that if it is $8°C$ inside it is possibly $0°C$ outside.

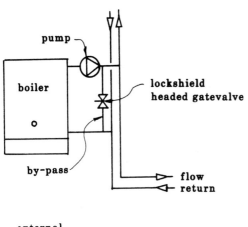

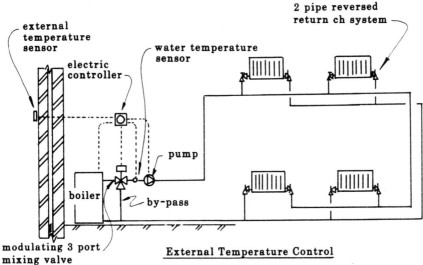

External Temperature Control

The flow temperature is continuously & automatically
adjusted to compensate for changes in outside temperature

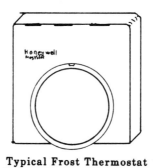

Typical Frost Thermostat

Design Concepts in Central Heating

Warm Air Heating

Relevant British Standard
BS 5864

A design of c.h. in which warm air is blown through a system of ductwork to the rooms to be heated, assisted by a fan. The air in this system is either heated directly in a special boiler (in which the air circulates around the combustion chamber), or the air is heated indirectly, in which case the air passes over a water-filled heat exchanger.

The direct system tends to be faster in its heating-up period, owing to the fact that the air is heated directly and no heat will have been lost from the flow and return pipe to the warm air heat exchanger. With this system, because a flue is required, it is not always possible to site the boiler in a desired central position; with the indirect system, however, the boiler can be sited away from the warm air heat exchanger (see figure).

The indirect system is generally more expensive to install but has the advantage of also heating water for domestic purposes, the boiler used being the same as that in a water-filled system of central heating. Special boilers can, however, be purchased for the direct system which will allow for water to circulate through the boiler for dhw purposes.

Once the air has been heated it is passed through a system of ductwork. The delivery air temperature at the room register/diffuser should not exceed 60°C.

Air from the heated rooms should be returned to the heater for reheating. If possible, the return air grill should be positioned opposite the warm air inlet diffuser into the room; for example, should the warm air inlet be at low level on one side of the room the return air outlet should be located on the opposite wall at high level. No return air grills are positioned in bathrooms, kitchens or WC compartments because of the large amounts of condensation and possible odours which can be drawn into the system. The return air is either passed back to the heater by a system of ductwork or the return air from the rooms is drawn back to the hall and eventually through a duct in the heater.

Should the second method be chosen, allowing the return air to flow to the hall, the air is simply allowed to flow through the grills in the internal walls; note that some thought must be given to their siting in bedrooms, etc., for privacy. To ensure good comfortable room conditions, up to 25% fresh air is often mixed with the return air: the heater air inlet manifold is fitted with a damper arrangement to regulate the proportions of fresh and return air. In large industrial type buildings often only fresh air is used from outside the building and no return air used at all.

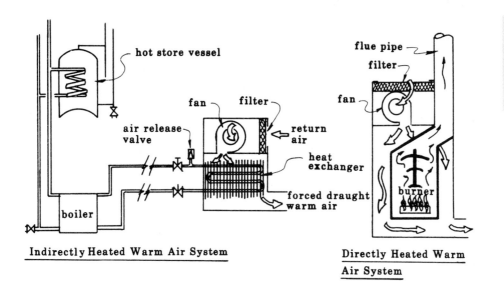

Indirectly Heated Warm Air System

Directly Heated Warm Air System

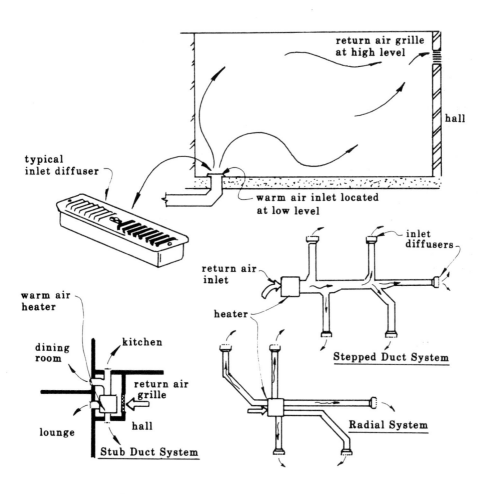

Warm Air Heating

Commissioning of Wet Central Heating Systems

Relevant British Standard
BS 5449

On completion of a c.h installation the system should be commissioned as follows:

(1) The pump should be removed and replaced with a suitable piece of pipe to bridge the gap. The system is now filled with water, any air being vented out as necessary, from all high points and the system checked for any leaks.

(2) The system should be drained out, and should receive a flush through to remove any wire wool, etc. The pump is now replaced and the system refilled, as in (1) above. The boiler is now made to ignite and the system brought into operation by turning up any thermostats, etc. A check should be made of the boiler thermostat to confirm that it is working.

(3) At this stage the boiler can be commissioned for correct operation. See pages 186 and 214 commissioning of gas and oil appliances.

(4) Close all lockshield valves and go round the system balancing the heat emitters to each room. This involves slowly reopening the lockshield valve to give a mean water temperature to each heater. Ensure that the design temperature drop across the system is maintained at the boiler; this should be carried out using clamp-on thermostats at the flow and return pipes to achieve approximately $82°C$ flow and $71°C$ return temperatures. This will ensure maximum efficiency and give a longer life to the boiler.

(5) Check the operation of the programmer, room and cylinder thermostats, etc., and any motorised valves fitted to the system and ensure the pump switches off as required.

(6) Recheck for leaks. Turn off the boiler and drain the system while the system is still hot. This assists the removal of flux residuals.

(7) Refill and vent the system, adding an inhibitor, if applicable.

Having cured any problems, and secured cistern lids and insulation material, the installer should attach to the boiler a card identifying the date of installation and the name of the installer. Labels should also be fixed to any valves, etc., for identification purposes. The job should now be tidied up in readiness for handover to the owner.

(8) Handing over: The working of the system should be demonstrated to the user and the best methods of economic and efficient usage explained. All documentation supplied should be left with the owner/occupier who should be made aware of the need for a regular service contract to ensure that the equipment is maintained in an efficient and safe operating condition.

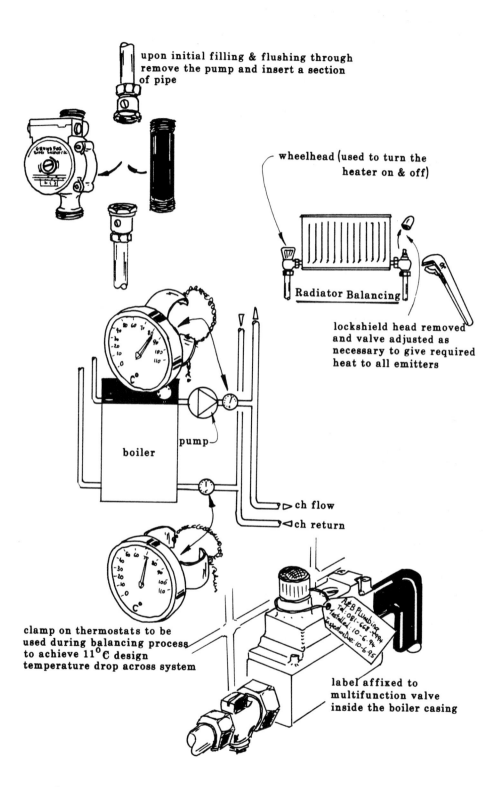

upon initial filling & flushing through remove the pump and insert a section of pipe

wheelhead (used to turn the heater on & off)

Radiator Balancing

lockshield head removed and valve adjusted as necessary to give required heat to all emitters

pump

boiler

ch flow

ch return

clamp on thermostats to be used during balancing process to achieve 11°C design temperature drop across system

label affixed to multifunction valve inside the boiler casing

A&B Plumbing
Tel: 081-668-1994
Installed: 10-6-94
Service Due: 10-6-95

Commissioning of Wet Central Heating Systems

Part
Gas Supplies

4

Properties and Combustion of Natural Gas

Properties of natural gas

Constituent	Chemical formulae	Approximate % by volume
Methane	CH_4	90.0
Ethane	C_2H_6	5.3
Propane	C_3H_8	1.0
Butane	C_4H_{10}	0.4
Carbon dioxide	CO_2	0.6
Nitrogen	N_2	2.7

Natural gas occurs below ground and is a by-product of the breaking down of decaying vegetation, etc., which is subsequently trapped beneath impervious land strata. The gas is non-toxic and odourless; the smell is added to the gas by the supplier to assist the identification of a leak. The relative density of natural gas is approximately 0.58, so it is lighter than air and will rise as a result. Natural gas is primarily made up of hydro-carbons which are a combination of hydrogen and carbon atoms. Both these atoms can be burnt in the presence of oxygen (O_2) and when burning their chemical composition will change. For example, if we looked at methane burning in a sufficient quantity of oxygen to give complete combustion, the following reactions would occur:

Methane (1 volume) CH_4	+	Sufficient oxygen (2 volumes) $2 O_2$	gives off	Carbon dioxide (1 volume) CO_2	and	Water vapour (2 volumes) $2 H_2O$

Note: none of the atoms is ignored. Nothing has actually been consumed; its chemical state has simply been changed. The two gases given off are completely harmless and safe to breath in, being present in the surrounding air. If we looked at the same volume of gas with an insufficient quantity of oxygen, incomplete combustion would result, as shown:-

Methane (1 volume) CH_4	+	Insufficient oxygen (1 volume) O_2	gives off	Water vapour (1 volume) H_2O	and	Unburnt fuel CO H_2

Apart from being inefficient, this situation is also highly dangerous, as the gas CO (carbon monoxide) is highly toxic; in fact, 0.4%/volume in the air can prove fatal within a few minutes.

There are two fundamental requirements for the ignition of gas: a supply of oxygen and a gas temperature of approximately $704°C$. The gas must be mixed with the oxygen to give a ratio of gas in the air of between 5% and 15%. Upon ignition the flame will burn through the gas/air mixture at a rate of 0.36 m/s. The gas burner manufacturer designs the burner to allow the gas to be pre-mixed with the air prior to leaving the burner head and turbulate in the reaction zone, allowing complete mixture and warming of the fuel.

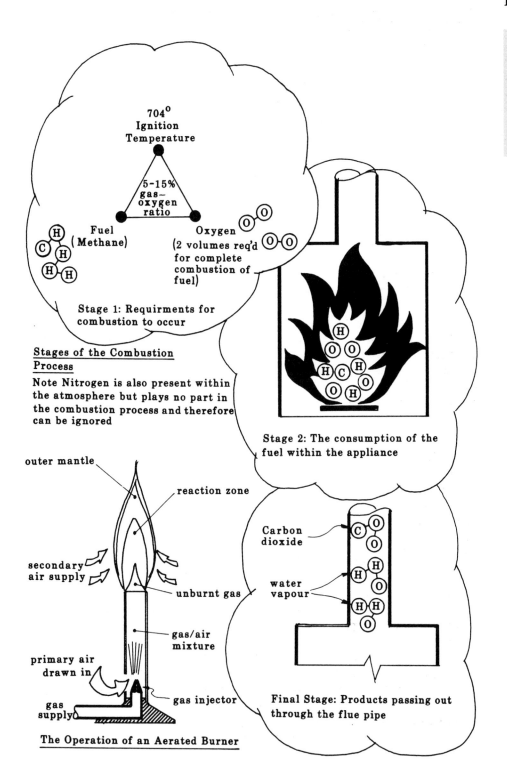

704°
Ignition
Temperature

5-15%
gas–
oxygen
ratio

Fuel
(Methane)

Oxygen
(2 volumes req'd
for complete
combustion of
fuel)

Stage 1: Requirments for
combustion to occur

Stages of the Combustion
Process

Note Nitrogen is also present within
the atmosphere but plays no part in
the combustion process and therefore
can be ignored

Stage 2: The consumption of the
fuel within the appliance

outer mantle

reaction zone

secondary
air supply

unburnt gas

Carbon
dioxide

water
vapour

gas/air
mixture

primary air
drawn in

gas
supply

gas injector

Final Stage: Products passing out
through the flue pipe

The Operation of an Aerated Burner

Properties and Combustion of Natural Gas

Liquefied Petroleum Gas Installations

Relevant British Standard
BS 5482

Many of the requirements for LPG are the same as those for natural gas; for example, the same ventilation and flue requirements must be met, as laid down in BS 5440. However, because LPG is not supplied from the national supplier, British Gas, from pipes in the road, and is delivered instead in pressurised liquid form via a bulk tank or cylinder, some special provisions need to be made.

The fundamental requirements for ignition to occur include a 1.9–8.5% gas-in-air mixture for butane and 2–11% gas-in-air mixture for propane, with an ignition temperature of 480–540°C. The relative densities of the fuels are 2.0 and 1.5 respectively; in both cases, therefore, the gases are heavier than air. LPG injectors are somewhat smaller than natural gas injectors.

Gas supply considerations Where cylinders are used to store the LPG they should be located outside the building and afford easy access. Butane is sometimes fitted inside domestic premises. The requirements of the Health and Safety Executive codes of practice should be observed. It is essential that all cylinders are stored upright with the valve at the top and it is essential that the stored supply is above ground level and never adjacent to open drains or basements where gas resulting from a leak might accumulate. Bulk storage tanks may also be fitted in accordance with the requirements of the Home Office and the gas supplier's specification.

LPG installations are divided primarily into high- and low-pressure stages, the high pressure being that in the cylinder and pipework up to the pressure regulator. Whenever possible, reserve cylinders should be connected to a manifold and fitted with non-return valves; this will permit one cylinder to be removed without the need to shut down the whole system. Guidance is given here as to the number of cylinders that should be provided in parallel to supply an installation. To ensure a continuity of supply it is advisable to double this capacity for replacement cylinders.

Example: for an installation consisting of a domestic cooker, boiler and small fire the minimum number of 14.5 kg butane cylinders should be 1.75 + 2.02 + 0.20 = 3.97, i.e. four; and where 19.0 kg propane cylinders are chosen the minimum number should be 0.88 + 1.02 + 0.12 = 2.02, i.e. 2.

Storage requirements – approximate No of LPG cylinders required

Appliance	Butane 14.5 kg	Propane 13.0 kg	19.0 kg	47.0 kg
Cooker	1.75	1.15	0.88	0.49
c.h. boiler	2.02	1.28	1.02	0.57
Small fire	0.20	0.15	0.12	0.07
Large fire	0.80	0.50	0.40	0.22
Multipoint	3.20	2.10	1.68	0.93
Single point	1.30	0.85	0.68	0.38

Testing the high pressure stage This is carried out by firstly closing off the low pressure line and then inserting a pressure gauge ranging from 0 to 10 bar between the cylinder valve and pressure regulator. The cylinder valve is opened to charge the pipe and reclosed. After a 1 min temperature stabilisation period there should be no drop in pressure over a test period of 5 min. For installations using non-return valves the procedure will need to be repeated for both sets of cylinders. After a successful test remove the pressure gauge and test this joint using leak detection fluid.

Testing the low pressure stage For new installations, prior to gas connection, the system is charged with air up to 45 mbar; after a 5 min temperature stabilisation period there should be no pressure drop over a further 5 min test period. (See page 162 for a guidance on the procedure.)

The maximum working pressure drop across the system when all burners are operating should not exceed 2.5 mbar (see notes on page 164, Pressure and Flow). Because no meter is generally fitted to LPG installations it is not possible to calculate the heat input to the burner as with natural gas; however, should a check be made on the combustion efficiency of the appliance (see page 184, Flue Efficiency), an indication will be given of its correct input in accordance with the manufacturer's data.

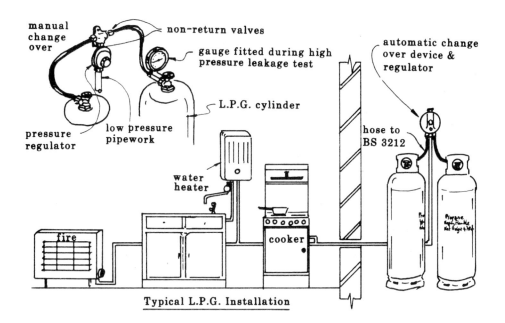

Typical L.P.G. Installation

Liquefied Petroleum Gas Installations

Gas Safety Requirements

Any person carrying out any work in relation to gas fitting must be competent to do so. The Gas Safety (Installation and Use) (Amendment) Regulations 1990 clearly define what can and cannot be done.

You may do the work yourself, in your home, providing you comply with the Regulations, but if you wish to do the work for gain (i.e. be employed by someone) you must be a member of a class of persons approved by the Health and Safety Executive (Regulation 3.3). The Council for Registered Gas Installers (CORGI) operates such a scheme. To simplify this situation, if you are an employee, your employer must be registered and you must be a named operative with the company. If you work for yourself then you yourself must register direct with CORGI.

Over 50 Regulations are identified and many refer to other documents such as British Standards, Building Regulations and Manufacturers' data, etc., all of which must be complied with. It would not be possible for a book such as this to identify all the Regulations which need to be observed but the following is a brief description of just a few from the Gas Safety (Installation and Use) (Amendment) Regulations 1990:

Regulation No 7 No person shall alter the building in any way so as to affect the safe operation of the gas fitting. Builders and double glazing people, etc., may inevitably break this law, but are still liable to prosecution.

Regulation No 9 Where any pipework is to be removed a suitable cross bonding wire must be incorporated to prevent the production of a spark or electric shock due to a fault in the electrical supply (see Part 6 Electrical Work, page 217).

Regulation No 18 No pipes should be run in a cavity wall, except when passing from one side to the other. Where this is the case a sleeve should be incorporated and sealed to prevent the passage of gas. No pipe must be installed in an unventilated shaft, duct or void.

Regulation No 29 The installer must leave with the owner/occupier all instructions provided by the manufacturer.

Regulation No 30 Open flued appliances are not to be installed in private garages or bath/shower rooms. In addition, bedrooms should also be avoided to satisfy BS 6798 and CORGI requirements.

Regulation No 33 When an appliance has been connected to the gas supply it must be tested in accordance with the manufacturer's instructions (i.e. commissioned).

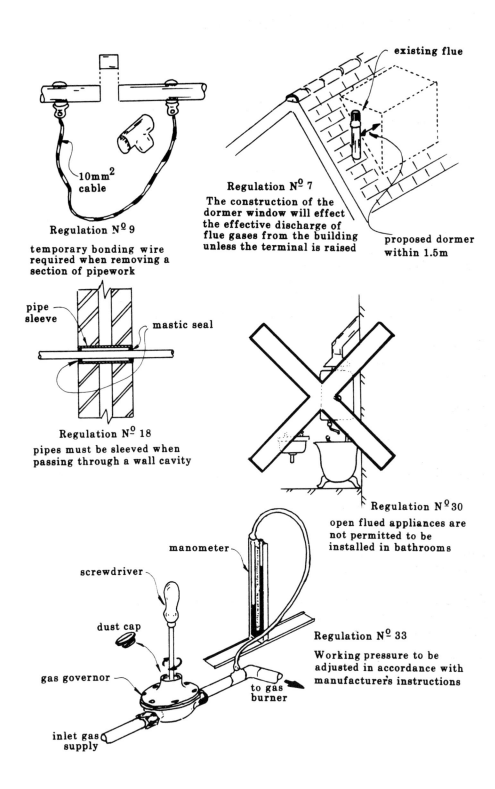

Regulation N⁰ 9

temporary bonding wire required when removing a section of pipework

Regulation N⁰ 7

The construction of the dormer window will effect the effective discharge of flue gases from the building unless the terminal is raised

proposed dormer within 1.5m

existing flue

Regulation N⁰ 18

pipes must be sleeved when passing through a wall cavity

Regulation N⁰ 30

open flued appliances are not permitted to be installed in bathrooms

Regulation N⁰ 33

Working pressure to be adjusted in accordance with manufacturer's instructions

Gas Safety Requirements

Gas Supply to the Consumer

Relevant British Standards
BS 6400 and BS 6891

When a supply of natural gas is required, provided it is available in the local district, a service pipe is run to the building (see figure). Whenever possible, a special meter box should be installed. Three designs will be found: the semi-concealed, the surface-mounted or the sunken box. The meter may be located inside a garage or the dwelling itself, although these are locations which are not recommended because the supplier will have difficulty in obtaining meter readings.

The service pipe should not be routed below the foundations or through unventilated voids, and where the service pipe has to pass through a wall the pipe must be sleeved to allow movement. The sleeve should be made good with mortar in the wall and the gap between the pipe and sleeve sealed with a mastic sealant to prevent the passage of gas.

The supply gas pressure may be as high as 75 mbar, which is far in excess of that required by the consumer and is therefore reduced to a supply pressure of 20 mbar, and in no cases exceeds 25 mbar. This is achieved by the installation of a meter governor, which adjusts automatically to provide an adequate supply of gas at a constant pressure. The meter governor is set by the supplier of the gas and must not be adjusted, indicated by a lead seal, any excessive or insufficient pressures being brought to the attention of the supplier.

The main consumers' emergency control valve must be located in a readily accessible position and have a securely attached lever arm. When the valve is located in a vertical pipeline it should be closed when the lever has been moved as far as possible down-wards and in a horizontal plane. A label must be included to show the open and closed positions of the valve (see Gas Safety Regulation No 8).

Internal shafts

Where a continuous shaft is incorporated to accommodate multi-storey properties, the shaft must be constructed adjacent to, or within 2 m of, an external wall and be ventilated at high and low level to the external environment, thus preventing any build-up of gas in the event of a leak. The individual branches to each apartment are taken off the riser, to include a service valve, and taken up vertically, to incorporate an expansion-type coupling, before turning horizontally through the shaft wall into the room, thus allowing for movement without fracture.

The horizontal pipe passing into the room must be fire-stopped as necessary, to prevent the passage of gas or smoke. The vertical shaft may be either one long continuous shaft or fire-stopped at each level, depending upon the building design and local Regulations, but in all cases the shaft must be ventilated in the way described above.

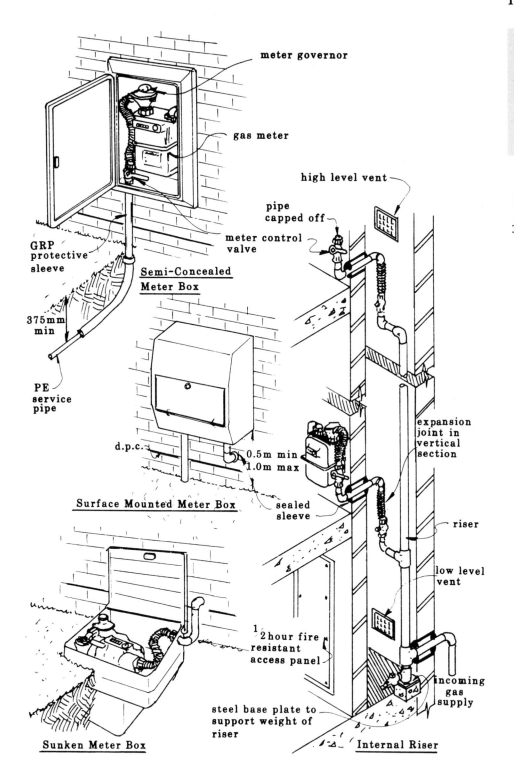

meter governor

gas meter

high level vent

pipe capped off

meter control valve

GRP protective sleeve

Semi-Concealed Meter Box

375mm min

PE service pipe

d.p.c.

0.5m min
1.0m max

Surface Mounted Meter Box

sealed sleeve

expansion joint in vertical section

riser

low level vent

$\frac{1}{2}$ hour fire resistant access panel

steel base plate to support weight of riser

incoming gas supply

Sunken Meter Box

Internal Riser

Gas Supply to the Consumer

Pipe Sizing of Gas Pipework

Relevant British Standard
BS 6891

When pipe sizing hot or cold water pipework, if the pipework is undersized the worst that can happen is that an appliance becomes starved of water flow, which results in inconvenience and inefficient usage. There is no danger to the user or occupier of the building. With gas supply pipework on the other hand, a dangerous situation may result from undersized pipework. If the pressure is insufficient at the burner not enough primary air will be drawn in to achieve complete combustion, which may result in the production of carbon monoxide, or in very bad cases a flash-back into the pipework may occur which could lead to an explosion. The maximum pressure drop between the meter and the furthest appliance, under maximum flow conditions, must not exceed 1 mbar.

Note: unfortunately gas meters still register in cubic feet (ft^3); therefore, when pipe sizing I prefer to work in British Imperial Units, which gives the gas rate in ft^3; one can, of course, employ metric equivalents by using conversion factors or tables.

Shown opposite is a completed example of a gas carcass which has been sized in stages and is achieved as follows:

Column 1 This is the section of pipework which is being sized; note that the gas carcass has been broken down into various sections.

Column 2 The actual length of the pipe is measured and inserted here.

Column 3 An allowance must be made for frictional resistance through fittings. Each time the gas passes round an elbow or tee fitting 0.5 m should be allowed, and 0.3 m in the case of pulled 90° bends.

Example: in section A three elbows are used and one tee fitting; therefore: $4 \times 0.5 = 2$ m.

Column 4 This is the sum total of columns 2 and 3 (thus in section A, 4.5 m + 2 m = 6.5 m), this being the effective length resulting from the actual length and additional length due to frictional resistance.

Column 5 This is found by adding the total btu/h rating for all appliances being supplied by the section in question and dividing this figure by 1000 to convert to gas consumption in ft^3.

Example: section A =

boiler:	58 000	btu/h
cooker:	50 000	btu/h +
fire:	17 000	btu/h
	125 000	

Therefore   125 000 ÷ 1000 = 125 ft^3

Gas carcass serving boiler: cooker and fire

	1	2	3	4	5	6	7	8	9	
	Section	Measured length	Additional length due to fittings	Effective length	Required gas flow rate	Suggested pipe size	Maximum length allowed	Pressure loss	Progressive pressure loss	Notes
		(m)	(m)	(m)	(ft³)	(mm)	(m)	(mbar)	(mbar)	
	A	4.5	2.0	6.5	125	22	12	0.5	0.5	✔
	B	3.6	1.0	4.6	17	15	30	0.2	0.7	✔
	C	4.0	0.5	4.5	108	22	15	0.3	0.8	✔
	D	2.0	1.0	3.0	50	15	9	0.3	1.1	Pipe undersized
	D	2.0	1.0	3.0	50	22	30	0.1	0.9	✔
	E	0.6	0.5	1.1	58	15	6	0.2	1.0	✔

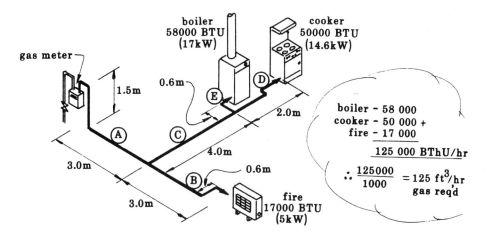

Pipe Sizing of Gas Pipework

Column 6 At this stage a pipe size is suggested which the table will confirm (or not) for possible use.

Column 7 By referral to Table 1, opposite, the pipe diameter is aligned against the column showing the length of pipe through which the required gas flow will pass.

Example: in section A, where a 22 mm pipe was suggested and a flow of 125 ft^3 required, the maximum length of pipe run is indicated to be 12 m, through which 140 ft^3 of gas could pass, whereas only 120 ft^3 of gas could pass through the next column, i.e. 15 m.

Column 8 The pressure loss for the section in question is found by dividing column 4 by column 7; thus, in section A $6.4 \div 12 = 0.5$.

Column 9 The progressive pressure loss is the sum total of the pressure losses for each section preceding the section in question. In section A the progressive loss is the same as the pressure loss, but in section E, for example, the progressive pressure loss will be the sum total of sections A, C and E; therefore: $0.5 + 0.3 + 0.2 = 1.0$ which, in this example, will be of sufficient size.

In conclusion, one estimates a suggested pipe diameter and completes the table for the section in question to prove its suitability for use or not. If it proves undersized one simply goes back to column 6 to choose a larger pipe diameter. It may be that one has to increase the diameter of the first section, as in our example 0.5 mbar was generated prior to the first tee connection. Sometimes one opts for larger pipework to allow for possible extensions to the gas pipework in future years, e.g. a larger boiler or extra gas fire.

Meter sizing

In completing this work on pipe sizing one can see the meter size required, which in our example suggested we needed 125 ft^3; therefore, the smallest domestic meter would have been sufficient, which supplies up to 212 ft^3/h. This capacity is indicated on the meter face plate.

Table 1 Flow discharge of natural gas, in ft³, from Table X (BS 2871) copper tube with a 1.0 mbar pressure differential between each end

Pipe diameter (mm)	Discharge of gas flow rate (ft³)							
	Length of pipe in metres							
	3	6	9	12	15	20	25	30
10	30	20	18	13	11	8	6	5
12	54	36	30	29	24	18	14	12
15	100	69	54	45	40	34	32	31
22	310	210	160	140	120	100	89	80
28	630	420	330	280	250	210	180	170

Table 2 Flow discharge of natural gas, in ft³, from medium gauge (BS 1387) steel tube with a 1.0 mbar pressure differential between each end

Pipe diameter (in)	Discharge of gas flow rate (ft³)							
	Length of pipe in metres							
	3	6	9	12	15	20	25	30
¼	28	19	17	13	10	8	6	5
⅜	73	49	39	33	30	29	24	20
½	150	100	82	70	61	52	46	44
¾	340	230	190	160	140	120	100	93
1	650	440	350	300	260	220	200	180

(Tables above reproduced with permission from BS 6891 *Specification for installation of low pressure gas pipework of up to 28 mm (R1) in domestic premises (2nd family gas).)*

4 Gas Supplies

Testing for Soundness and Purging

Relevant British Standard
BS 6891

Pressure tests are carried out on gas installations to check for leaks. Two types of test may be carried out, these being for new or existing installations. Should the test not hold to the pressure required a leak is present and should be found using a leak detection solution. As the liquid is washed around the joints it will bubble up, provided a small pressure is maintained within the pipeline.

New installations prior to connection of meter

(1) All open pipe ends should be capped off except one onto which is fitted a testing tee, comprising a small plug cock and a pressure test nipple.
(2) A manometer, filled with water to register zero, is connected to the test nipple and air is pumped or blown into the system to register 30 mbar (300 mm). This will give a test pressure of one and a half times the system pressure and ensures soundness; however, BS 6891 suggests only a 20 mbar test.
(3) The plug cock is shut off and there is a wait of 1 min for the air to stabilise.
(4) After the initial 1 min a further 2 min is waited in which there should be no further pressure drop at the manometer.

Existing installations (using the gas pressure from the service main itself)

(1) Turn off the main gas inlet control valve and all gas appliances and pilot lights.
(2) Remove the screw from the test nipple located on the outlet side of the gas meter and connect a manometer to it; adjust the gauge to zero.
(3) Slowly open the main gas cock to allow 2–4 mbar through, then re-shut the valve. There should now be no slow rise in the pressure reading at the gauge. If there is, it indicates that the main gas cock is letting gas through when closed; in these circumstances it is impossible to continue the test to achieve a true reading and the gas authority should be notified.
(4) Assuming the main gas cock does not let-by, **slowly** reopen the valve to allow gas to enter the pipework and come up to the standing pressure of the system. This should be between 20 and 25 mbar; where it is not the supplier of the gas should be informed.
(5) Turn off the gas and wait 1 min for stabilisation; then, over the following 2 min, the pressure drop registered on the manometer should show no further drop. Where appliances are connected but the operating taps and pilots are simply turned off, a permissible pressure drop is allowed and must not exceed 4 mbar in domestic situations, using a meter with a capacity up to 0.071 ft^3/rev. (See BS 6891, which deals with larger meters.)
(6) Upon the test proving satisfactory the test nipple is replaced and the gas turned on. Leak detection spray is now applied to the test nipple connection and any pipework preceding the meter. Provided there is no smell of gas the system can be regarded as sound.

Purging

Prior to gas entering the pipework, air which is trapped needs to be expelled. This is referred to as purging. Purging also removes debris, such as wire wool, from the pipeline. During any purging operations one must ensure good ventilation, prohibit naked flames and avoid the operation of electric switches. The procedure is carried out as follows:

(1) Turn off the emergency control valve at the meter and disconnect the supply pipe at the furthest appliance.
(2) Open the control valve and allow not less than five times the badged capacity/rev of gas through the meter (for the domestic meter with a capacity of 0.071 ft^3/rev this will be $0.071 \times 5 = 0.36$ ft^3): therefore three and a half divisions on the test dial. A smell of gas must also be evident at the disconnected joint.
(3) Reconnect the appliance and check the soundness of the joint broken. Any further legs of pipe will only need to be opened until gas is smelt and is freely discharging from the open ends.

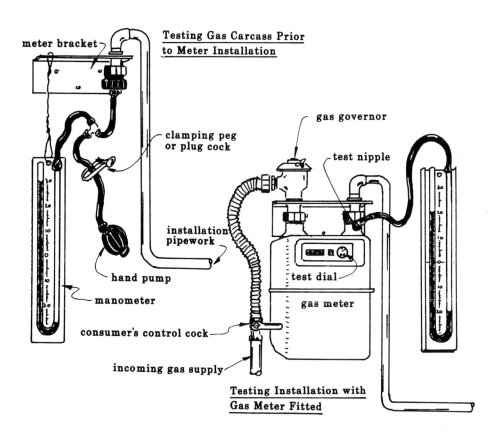

Testing Gas Carcass Prior to Meter Installation

meter bracket

gas governor

clamping peg or plug cock

test nipple

installation pipework

hand pump

manometer

test dial

gas meter

consumer's control cock

incoming gas supply

Testing Installation with Gas Meter Fitted

Testing for Soundness and Purging

Pressure and Flow

Gas governors

These are devices which are fitted into the pipeline to ensure that the gas arriving at the appliance does so at a constant pressure, as recommended by the manufacturer. Modern governors use a spring to act upon a diaphragm exerting a force; when the gas exerts a force on the underside of the diaphragm in excess of that of the spring, it causes the diaphragm to lift and carry with it the valve, thus reducing the gap through which the gas can pass to the appliance.

Should the gas pressure drop, the spring forces the diaphragm down again, increasing the gap and allowing more gas through. Conversely, should the pressure increase the diaphragm is lifted higher, closing the valve gap even more. This continuous up-and-down movement of the valve ensures that a balance is maintained between the inlet and outlet to the governor. The pressure acting on the spring can be adjusted by means of a screw on top of the governor.

Standing pressure This is the pressure available for use with no appliances *on*. The standing pressure at the meter should be 20–25 mbar; where it is not within this range the supplier should be notified. The standing pressure can be taken anywhere throughout the system, from a suitable test point, on the gas carcass side of any governing devices.

Working pressure at appliances This is the pressure recommended by the manufacturer and is adjusted while the appliance is burning fuel. Prior to lighting up any burners a manometer is positioned on the test nipple of the appliance side of the governor. The burner is ignited and left to run for some 5 to 10 minutes, thus allowing the pressure to stabilise. The pressure-adjusting screw on the governor is then turned as necessary to increase or decrease the pressure; it is turned clockwise to increase the pressure.

Working pressure of the system

This test is one which ensures that the maximum pressure drop across any gas supply installation does not exceed 1 mbar (as identified under pipe sizing *see* page 158). The test is carried out as follows.

One manometer is positioned at the meter and a second one at the furthest appliance. All burners on the system are ignited and after a suitable stabilisation period (5–10 min) a reading is taken from each manometer, the difference between the two being the system working pressure drop. Where only one manometer is available the test is done in two stages.

Heat input (gas rate)

The amount of gas supplying a burner is not only dependent upon the pressure but also on the injector size used; for example, the gas injector for LPG is much smaller than that used for natural gas. To ensure that the correct volume of gas is being consumed as recommended, i.e. to ensure that the correct injector is fitted or that there is no blockage in the injector, one must check the gas input as follows:

(1) Turn on the burner and wait for the gas to stabilise
(2) Observe the test dial on the meter and record the time, in seconds, it takes to complete one complete revolution of the test dial (i.e. 1 ft³)
(3) Divide this figure into the number of seconds in 1 hour (3600)
(4) Finally multiply by 1000 to convert ft³ to btu/h.

Example: assume it takes 57 s for the gas passing to the burner to consume 1 ft³ of gas.

Therefore $\dfrac{3600}{57} = 63.158 \times 1000 = 63158$ btu/h

Where the heat input is required in kW multiply by the conversion factor of 0.0002931; thus: $63158 \times 0.0002931 = 18\frac{1}{2}$ kW

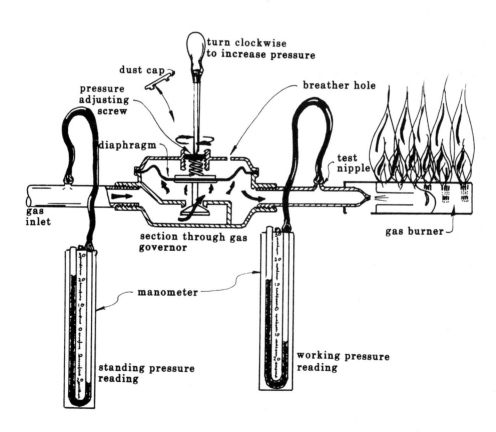

Pressure and Flow

Flame Failure Devices

This is a device, incorporated in the gas control valve of a burner, which prevents gas discharging from the main burner ports until the pilot flame has become established. There are several methods of flame detection, including:

Thermo-electric valve A device which converts heat energy into electrical energy. When heat is allowed to impinge on to a thermocouple an electric current is generated; this is used to operate a solenoid valve (an electromagnet). The valve works as follows:

(1) Gas enters the multifunction valve and passes through the governor to reduce its pressure to that acceptable for the boiler.
(2) The large button on the valve is depressed and held down; this opens the thermo-electric valve and allows gas through to the pilot jet. Simultaneously the smaller button is depressed several times; this operates a spark to ignite the pilot flame.
(3) Once the pilot has been lit for 16–20 s the flame playing on to the thermocouple energises the electromagnet thus holding valve 'X' open; the pressure can now be released from the large button.
(4) The main flow of gas can now continue to the next solenoid valve which only opens and allows gas to the main burners when the boiler thermostat is turned up and calls for heat, thus electrically energising this valve and causing it to open. Should the pilot be extinguished the thermocouple electromagnet will become de-energised.

Liquid vapour device This device is best illustrated in the figure and works as follows:

(1) When the thermostat calls for heat the solenoid valve is electrically energised and allows the gas to pass only to the pilot line and thus escape through its injector.
(2) The pilot flame is simultaneously lit by an automatic spark which is generated at its point of escape. When the pilot is established it is detected by a light sensor and the spark switches off.
(3) The flame impinges on to a sensor, filled with a volatile fluid, which expands and causes the bellows to push down on the lever arm opening the main gas valve.
(4) When the temperature has reached that required by the thermostat it de-energises the solenoid valve, shutting down the flow of gas.

Direct electrical control A method which detects the flow of electrons through the pilot flame itself, which in turn makes the circuit to a solenoid. Basically this system works as follows:

(1) When the thermostat calls for heat it energises a solenoid which allows gas to pass to the pilot flame, where the flame is established by a spark.
(2) The established pilot flame allows the electron flow of current to pass through the flame, thus making the electrical circuit to a relay valve.
(3) The relay draws in the iron rod which in turn makes the electrical circuit to the main gas solenoid, causing it to open. Should the pilot be extinguished the electron flow will cease and the electrical circuit is broken, closing off the gas supply.

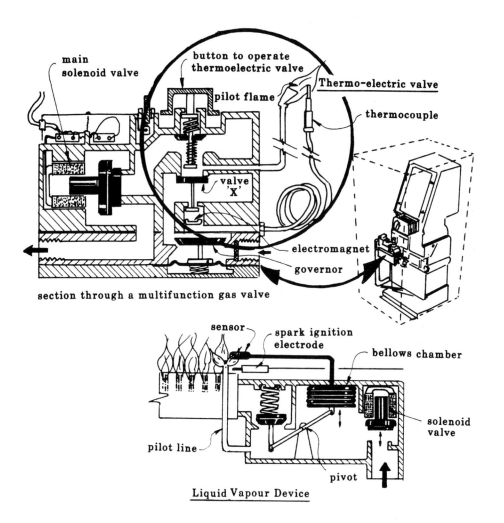

main
solenoid valve

button to operate
thermoelectric valve

pilot flame

Thermo-electric valve

thermocouple

valve
'X'

electromagnet

governor

section through a multifunction gas valve

sensor

spark ignition
electrode

bellows chamber

solenoid
valve

pilot line

pivot

Liquid Vapour Device

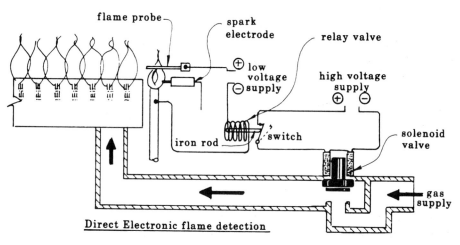

flame probe

spark
electrode

relay valve

$\oplus$ low
voltage
$\ominus$ supply

high voltage
supply
$\oplus$ $\ominus$

solenoid
valve

iron rod

switch

gas
supply

Direct Electronic flame detection

Flame Failure Devices

Open Flued Appliances (Conventional Flue)

Relevant British Standard
BS 5440

When installing an open flued appliance one must consider the location of the terminal and the route and length of the flue and its size. Failure to do so may result in excessive condensation problems or, more seriously, the inadequate removal of flue gases; this will mean an inefficient and dangerous flueing system, possibly leading to the death of the building occupants from carbon monoxide poisoning. The route which the flue takes on its way up through the building should be such that the flue gases are expelled as quickly as possible, thus ensuring the gases are not cooled to the due point of water (approximately 60°C) when condensation will occur.

Generally, increasing the height of the flue increases the flue draught because a high column of heated combustion products has a greater energy potential. However, surface friction also slows the flow; therefore the height cannot be indefinite; also, increased height increases heat losses. A typical height would be 6–8 m. Internal flues are far less likely to suffer the problem of cooling the gases; where exposure is likely insulate the flue or use twin-walled flue pipe. The size of the flue or its cross-sectional area should not be too small as spillage will occur and if it is too large greater heat loss will occur.

The driving force to remove the flue gases is due to either thermal convection currents (natural draught) or electric fan (forced draught). Systems which incorporate a fan have several advantages over natural draught systems including the increased positive removal of flue products, and greater flexibility in terminal location and siting of appliances. Alternatively potential disadvantages include possible noises from the fan, the tendency to depressurise the room containing the appliance and the need for additional safety devices to ensure the gas is shut down if there is a fan failure.

A draught diverter is incorporated with the appliance to prevent excessive down- or updraught due to adverse weather conditions; it also serves the purpose of diluting the flue gases. If the diverter does not form an integral part of the appliance and is fitted separately it must be installed within the same room or enclosure as the appliance. One should aim to have as much vertical flue as possible directly above the draught diverter (a minimum of 600 mm is recommended); this helps avoid the initial spillage of combustion products into the room.

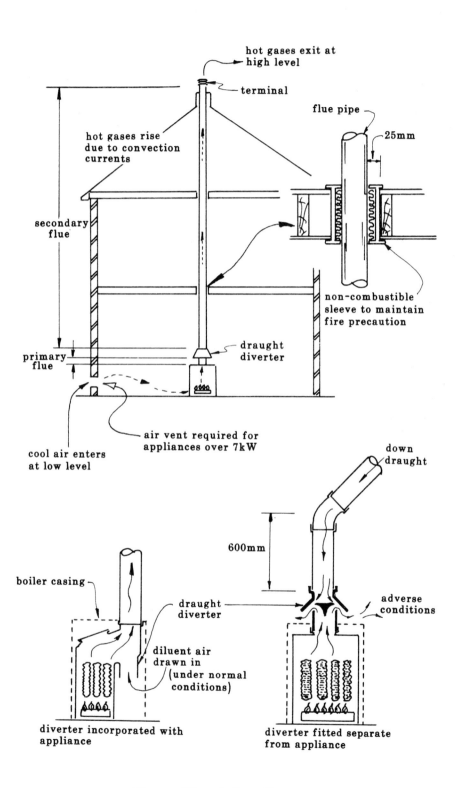

hot gases exit at high level

terminal

flue pipe

25mm

hot gases rise due to convection currents

secondary flue

non-combustible sleeve to maintain fire precaution

primary flue

draught diverter

air vent required for appliances over 7kW

cool air enters at low level

down draught

boiler casing

600mm

draught diverter

adverse conditions

diluent air drawn in (under normal conditions)

diverter incorporated with appliance

diverter fitted separate from appliance

Open Flued Appliances

Terminal Location for Open Flues

Relevant British Standard
BS 5440

Positive and negative pressure zones When wind blows towards a structure it creates a positive force (+) on one side of the building and a negative force (−) on the other. Where the wind has to travel over high structures the direction of the wind force is also caused to turbulate, blowing round in circles, as in a vortex. In each case these positive and negative pressure zones cause a blowing or sucking effect within their region. In the siting of any open flue terminal where the gases are being expelled by natural convection currents it is essential to discharge the products above these zones, otherwise excessive up or down draughts will occur, affecting the correct operation of the appliance.

Natural draught systems The terminal should provide the extraction of flue gases under virtually all wind conditions. The ideal position is above the highest point of the roof (e.g. ridge) and not shielded by any other structures or objects which might cause a pressure zone. Providing the location of the terminal is not within 1.5 m of a vertical surface such as a chimney stack or dormer window the height of the flue should be as indicated in the following table.

Recommended location of flue terminal outlet

Roof design	Minimum height between roof line and base of terminal (mm)
Flat roof without parapet	250
Flat roof with parapet	600*
Pitched roof up to 45°	600 or at or above the ridge
Pitched roof exceeding 45°	1000 or at or above the ridge

*Should the flue outlet be a greater horizontal distance than ten times the height of the parapet the height can be reduced to 250 mm

Where the terminal is within 1.5 m of a vertical surface the height of the flue needs to be at least 600 mm above the top of the structure. Note carefully from the example in the figure that the height indicated is from the highest point of the roof/flue intersection to the base of the terminal.

Forced draught systems With fan-assisted systems the flue route and final siting of the terminal is not critical, with respect to the performance of the appliance. However, positions where the combustion products may cause a nuisance should be avoided. Where the flue pipe terminates with a grille built into the wall the effective free area opening of the outlet should be at least 70% of the area of the flue pipe.

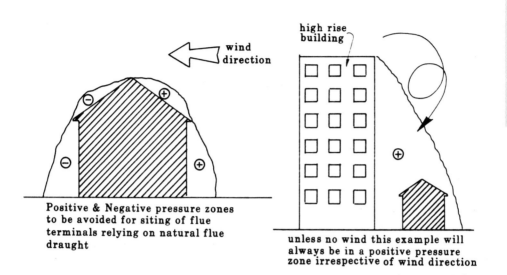

Positive & Negative pressure zones to be avoided for siting of flue terminals relying on natural flue draught

unless no wind this example will always be in a positive pressure zone irrespective of wind direction

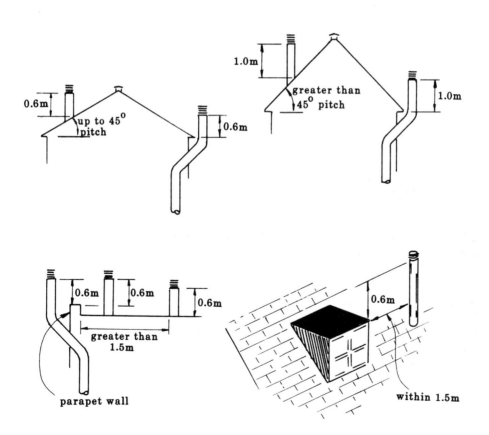

Terminal Location for Open Flues

Materials and Construction of
Open Flues

Brick chimneys Modern chimneys are lined with earthenware pipes during their construction and therefore should require no treatment. However, traditional 225 × 225 mm brick chimneys, designed for open fires, tend to be unnecessarily large for domestic gas burning appliances and as a consequence are prone to condensation problems; this inevitably leads to moisture passing into and through the brickwork. Generally speaking gas fires and gas circulators do not require the lining of the flue-way whereas boilers, having a greater heat input, do (see BS 5440).

Flexible stainless steel liners The size of the liner used should always be the same as, or larger than, the appliance flue outlet and is generally installed as follows:

(1) The existing chimney pot and its mortar flaunching is removed.
(2) A draw cord is lowered through the chimney and the liner attached to a dome-shaped bung and pulled up, or down if easier, through the flue way.
(3) The liner is fixed at the top, using a clamping plate, and at the base, again using a clamping plate or directly on to the appliance.
(4) The flue terminal is fitted and the flaunching made good, thus discharging the water away from the terminal. Note that for older buildings it is possible to purchase terracotta terminals, suitable for gas therefore being in keeping with the building design.
(5) The space between the base of the chimney and flue liner should be sealed (e.g. using glass fibre quilt) to create a void of trapped air and prevent flue gases passing up between the liner and the chimney.

Note that where the appliance is fitted externally to the chimney a flue pipe should be used to run up into the chimney void and the liner fixed to a socket at this point, this joint being available for future inspections via an access plate.

Precast-concrete flue blocks A specially designed flue system which is built into and forms part of the building structure, allowing for a larger living space. The system should incorporate a starter block at its base and, where possible, run vertically to its point of termination. Sometimes the system is only made up to the roof space and the final section run using twin-walled flue pipe terminating with a ridge terminal. Alternatively a brick chimney stack could be constructed through the roof section.

Flue pipes Large diameter pipes, available in a range of sizes and made from vitreous enamelled mild steel, stainless steel, high grade aluminium and asbestos type materials. Single- or double-walled pipe is available, the latter consisting of two concentric pipes separated by an air space or filled with an insulation material. When single-walled pipe is used with fire cement type joints the socket should sit below the spigot, thus ensuring that any water resulting from condensation will not cause the cement to weaken and fall out from the joint.

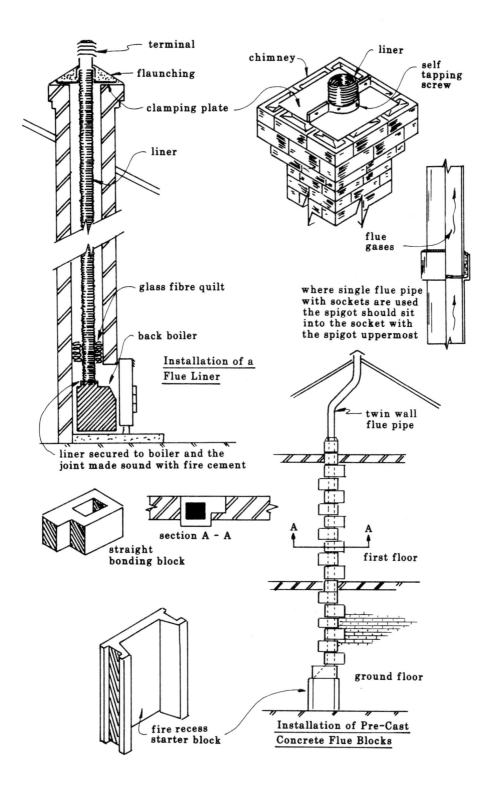

terminal

flaunching

clamping plate

liner

glass fibre quilt

back boiler

Installation of a
Flue Liner

liner secured to boiler and the
joint made sound with fire cement

chimney

liner

self
tapping
screw

flue
gases

where single flue pipe
with sockets are used
the spigot should sit
into the socket with
the spigot uppermost

twin wall
flue pipe

A   A

first floor

ground floor

straight
bonding block

section A - A

fire recess
starter block

Installation of Pre-Cast
Concrete Flue Blocks

Materials and Construction of Open Flues

Installation of Gas Fires

Relevant British Standard
BS 5871

Gas fires fall into three categories:

☐ *Traditional gas fires* These sit in front of the fireplace opening, making use of a closure plate.
☐ *Inset live fuel effect fires* These are either partially or fully inserted into the fireplace opening; the opening seal is usually incorporated with the fire.
☐ *Decorative fuel effect fires* These do not connect directly to the flue; the fire only sits in the grate within the fire opening. Such fires provide the room with very little radiant heat; they are primarily designed for decorative purposes (i.e. look like a real fire).

Connections to existing chimneys   When a gas fire is to be positioned into an existing masonry chimney which was previously used to burn some other fuel, the chimney must be swept thoroughly and all debris removed. The flue-way should be inspected for its sound condition and any dampers or restrictor plates removed or permanently fixed in the open position. A flue liner is not generally required, except where a large or long flue-way is encountered. It is not normally necessary to remove the existing chimney pot and fit a flue terminal.

When the chimney has satisfied its preliminary work it should be checked for a suitable up draught (i.e. pull) using a suitable smoke pellet. The chimney is firstly warmed with a blowlamp and the pellet inserted into the fire opening; all the smoke should be drawn up the flue-way.

No permanent ventilation is required for open-flued gas fires, providing they do not exceed 7 kW input. However, in the case of the decorative fuel effect (DFE) fires purpose-provided ventilation of at least 100 cm^2 will be required on all appliances up to 15 kW input.

Debris collection space When the gas fire is installed against the fireplace opening it should be noted that a 250 mm minimum void is to be maintained below the spigot entry point, thus allowing debris which may fall down the chimney to collect. When servicing any fire this space must be inspected.

Use of closure plate The traditional gas fire requires a closure plate (see figure) to prevent the entry of excess air into the fire whilst maintaining the correct gap to allow for suitable air relief resulting from excessive down draughts. The closure plate is secured in position using suitable adhesive tape, i.e. capable of maintaining its seal up to 100°C.

Fire precautions Floor standing fires are usually mounted on to a noncombustible hearth having a minimum thickness of 12 mm. The hearth should extend 300 mm forwards from the back edge of the fire and 150 mm beyond each edge of the naked flame or radiant. Wall-mounted fires should be installed so that the flame or radiant strips are at least 225 mm above any carpet or floor covering.

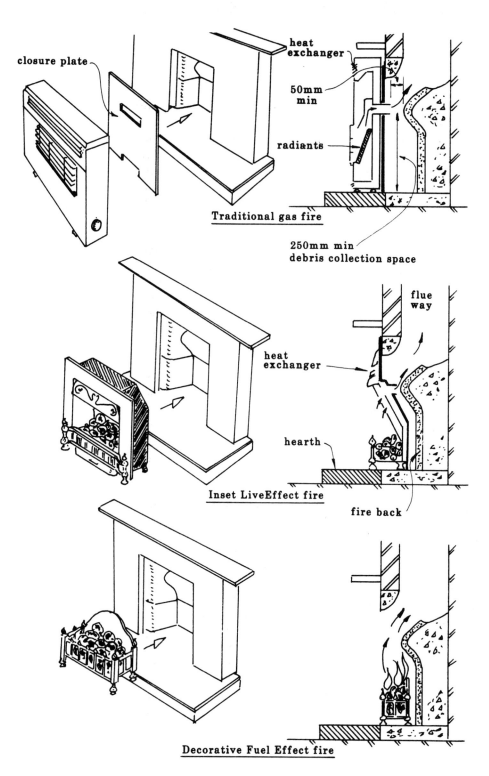

closure plate

heat exchanger

50mm min

radiants

Traditional gas fire

250mm min
debris collection space

flue way

heat exchanger

hearth

Inset LiveEffect fire

fire back

Decorative Fuel Effect fire

Installation of Gas Fires

Room Sealed Appliances (Balanced Flue)

Relevant British Standard
BS 5440

Room sealed appliances are ones which use a flue system which draws its air for combustion purposes from outside the building; these are unlike open-flued appliances which draw air from the room. Because of their design concept room sealed appliances should, whenever possible, be recommended as generally they present no danger to the occupants of the building from carbon monoxide poisoning (the result of vitiated air – air lacking oxygen – circulating between the room and appliance).

Natural draught systems In most cases appliances which expel the flue gases by convection currents (natural air flow) need to be located adjacent to an external wall. Natural draught appliances are easily identified by a typical large square terminal located on the external wall face.

Fan assisted systems These appliances have advantages over natural draught systems in that they can be located some distance from the external wall. Because the appliance is more positive in the extraction of flue gases, smaller outlet terminals are used; also, the system is much less affected by adverse wind conditions which result in the terminal not necessarily being located in a clear expanse of wall. The fan can be located on the air intake or the flue extract but in all cases the design of the burner should be such that, if the fan malfunctions, the burner will be prevented from firing.

Terminal location The location of the terminal should be such that the selected position ensures safe and efficient performance of the appliance. Unsatisfactory locations may result in:

- [ ] Flue gases entering the building through openings,
- [ ] Possible fire hazards,
- [ ] Staining to walls by the products of combustion,
- [ ] Inefficient combustion of fuel.

Additional points to consider include:

- [ ] The provision of a terminal guard where the flue outlet is less than 2 m above ground, balcony or flat roof level, where people have access,
- [ ] The provision of a suitable shield, at least 1 m long, to protect surfaces when a natural draught terminal is located within a distance of 1 m below a plastic gutter or within 0.5 m below any paintwork,
- [ ] An assurance that the combustion products do not cause a nuisance to passers-by or people on adjoining property.

Manufacturer's instructions should be sought for the correct location of the terminal; however, the table opposite can be taken as a general guide to suitable positions.

Suitable room-sealed terminal locations for gas appliances

Terminal position	Minimum distance (mm)	
	Natural draught	Fan assisted
Vertically from a terminal on the same wall	1500	1500
From an opening in a car port (e.g. window)	1200	1200
From a terminal facing a terminal	600	1200
From a surface facing a terminal	600	600
From an internal or external corner	600	300
Below balconies or projected roofs	600	200
Horizontally from a terminal on the same wall	300	300
Above ground or roof level	300	300
Below openings into the building (e.g. windows)	300	300
Below the eaves	300	200
Below gutters or discharge pipework	300	75
From vertical discharge pipework	75	75

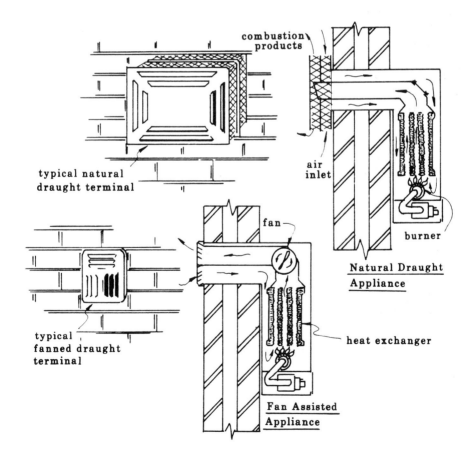

typical natural draught terminal

typical fanned draught terminal

combustion products

air inlet

burner

Natural Draught Appliance

fan

heat exchanger

Fan Assisted Appliance

Room Sealed Appliances

Ventilation Requirements

Relevant British Standard
BS 5440

Effective free area When installing an air grille it is essential to understand that a permanent unrestricted air flow is required; a closing device must not be incorporated with the grille and no fly screens attached, which may consequently become blocked. The effective free area is that through which the air may pass. The air vent should be direct to the outside air. Where this is not possible venting to an adjacent room is acceptable if the other room is provided with the correct size air vent.

Ventilation grilles to the outside environment may be located in any position (i.e. high or low) although when communicating between internal walls it should be located at low level (i.e. below 450 mm) to reduce the spread of smoke in the event of fire. The air supply requirements for an appliance will depend upon its location and heat input and can be identified as follows.

Room sealed appliances These do not require any air for combustion purposes.

Open flued appliances Appliances with an input below 7 kW do not require any additional air requirements, but where the input to the appliance is in excess of 7 kW the room in which the appliance is installed must have an air grille with a minimum effective free area of 450 mm^2 for every kilowatt in excess of 7 kW.

Example: A 22 kW open flued boiler requires a grille size of:

$$22 - 7 = 15$$
$$\text{therefore} \quad 15 \times 450 = 6750 \text{ mm}^2$$

Flueless appliances Appliances such as cookers may require an air grille; it depends upon the size of the room in which the appliance is installed. In every case an openable window, etc. direct to outside is required (see Table 1).

Table 1 Ventilation grille requirements for flueless appliances

Type of appliance	Maximum input (kW)	Room volume (m^3)			
		<5	5–10	10–20	>20
Domestic oven, hotplate or grill	N/A	10 000 mm^2	5 000 mm^2*	None	None
Instantaneous water heater	12	Not allowed	10 000 mm^2	5 000 mm^2	None

*Not required if a door opens directly to the outside.

Decorative fuel effect (DFE) fires (those which look like real fires) require an air grille of 10 000 mm^2 (see page 174, Installation of Gas Fires).

Multi-appliance installations Where a room contains more than one appliance the air vent requirement is based on the following:

(1) the total rated input for all open flued appliances, or

(2) the total rated input for all flueless appliances, or

(3) any larger requirements for any other type of appliance.

Air for cooling purposes Where a room-sealed or open-flued appliance is installed in a small cupboard, the compartment will require air vents for cooling purposes. Table 2 should be examined to identify the size of grille required.

Table 2 Air vent requirements for cooling (mm²/kW)

Type of appliance	Vented to:	High level grille (mm²)	Low level grille (mm²)
Room sealed	An adjacent room	900	900
Room sealed	The outside air	450	450
Open flued	An adjacent room	900	1800
Open flued	The outside air	450	900

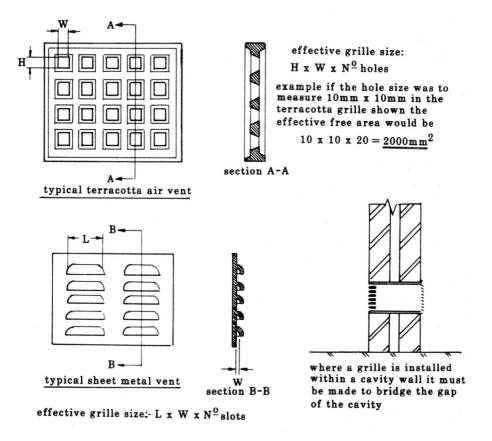

effective grille size:

H x W x Nº holes

example if the hole size was to measure 10mm x 10mm in the terracotta grille shown the effective free area would be

10 x 10 x 20 = 2000mm²

section A-A

typical terracotta air vent

typical sheet metal vent

section B-B

effective grille size:- L x W x Nº slots

where a grille is installed within a cavity wall it must be made to bridge the gap of the cavity

Ventilation Requirements

Other Flueing Systems

Relevant British Standard
BS 5440

Balanced compartments

An arrangement in which an open flued appliance is installed in a small room and the flueing and ventilation provisions are such that they in effect convert the operation of the appliance to that of a room-sealed appliance. For the system to work it is essential that the air intake and flue gas extract are located within the same positive or negative pressure zones. This is achieved by ensuring that the extract and intake terminals are within 150 mm of each other. For this system to function correctly it must not be influenced by pressures within the building; therefore, a self-closing, tight-fitting door must be provided with a suitable draught sealing strip included. Should the door be opened, the electrical supply to the appliance must be broken, thus ensuring it will not work. A notice should be fixed to the door saying it must be kept closed. **Note**: these compartments must not open into a bathroom. No openings should be allowed into the compartment other than those intended for air supply and flue gas extract (see figure). The air intake size is dependent on whether a high or low vent is provided in the compartment:

Ducted to high level: 2½ times 450 mm^2 for every kilowatt of the appliance.
Ducted to low level: 1½ times 450 mm^2 for every kilowatt of the appliance.

In each case this allows for cooling as well as for supplying air for combustion purposes.

Shared flue systems

Open flued appliances in the same room These are permitted to have their flue pipes joined together provided each appliance has its own draught diverter and flame failure device. Note that the air vent requirements are based upon the total heat input of all appliances fitted within the room.

Open flued appliances in separate rooms These can be connected into a branched flue system only if all the appliances are of the same type and the air supply for ventilation purposes is taken from an identical aspect. The subsidiary branches are designed to provide each appliance with its own pull, thus ensuring that the products are removed up into the main duct. The height of each branch should therefore be at least 3 m for gas fires and 1.2 m for other appliances. The main flue should not serve more than ten consecutive stories.

Room sealed appliances These can be connected to a vertical shaft (see figure); the combustion products rise by convection currents being set up, cool air entering at the base of the shaft via horizontal ducts or (in the case of buildings supported on columns) via air intake points, these being referred to as the Se-duct. In the case of a U duct system air enters via a second vertical shaft run down from roof level.

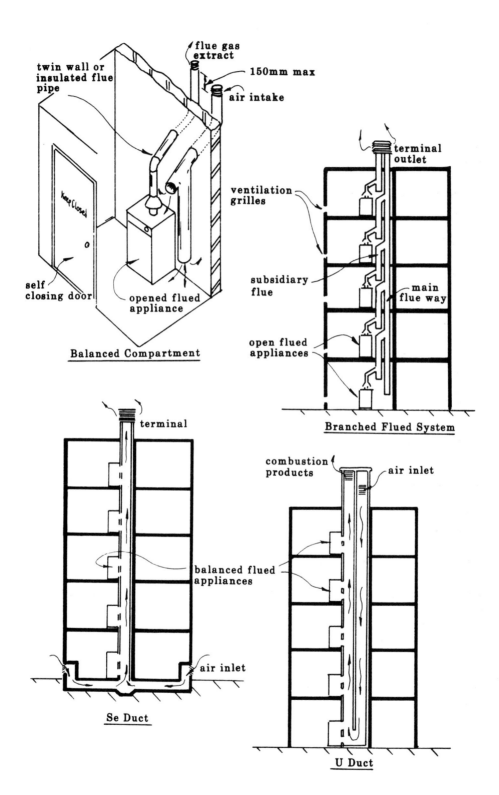

flue gas extract

twin wall or insulated flue pipe

150mm max

air intake

Keep Closed

ventilation grilles

self closing door

opened flued appliance

Balanced Compartment

terminal outlet

subsidiary flue

main flue way

open flued appliances

Branched Flued System

terminal

balanced flued appliances

air inlet

Se Duct

combustion products

air inlet

balanced flued appliances

U Duct

Other Flueing Systems

Combustion Analysis

Relevant British Standard
BS 1756

Sometimes one wishes to know the contents of the flue gases; this would assist, for example, in finding the efficiency of the appliance or when checking for the presence of carbon monoxide (CO). Several types of apparatus can be employed to assist one to find these results, ranging from electronic flue gas analysers to a simple suction apparatus such as the *Draeger* or *Fyrite* (see figures).

Draeger analyser

This apparatus is commonly used to find the concentration of CO in parts per million (ppm) although it can be used to find the concentration of any gas. It consists of a bellows unit into which is fitted a special glass sampling tube containing crystals capable of absorbing different gases, thus indicating their presence. When assembled (see figure) and the bellows operated as per the manufacturer's instructions, one can simply read from the scale on the tube. Should the concentration of CO in ppm need to be expressed as a percentage, it is simply divided by 10 000.

Fyrite analyser

This device is the same as that used to sample oil burner efficiencies. A special liquid capable of absorbing carbon dioxide (CO_2) is held in a container. When a sample of combustion products is forced into the analyser and the unit inverted, the gas is absorbed into the liquid; this causes the liquid to rise and allows one to read the percentage CO_2 from an indicator scale on the side of the container. See also page 212 on combustion efficiency testing for oil burners.

CO/CO_2 ratio To ensure the safe operation of any appliance burning natural gas it must be checked to contain a ratio of CO to CO_2 **not in excess** of 0.02. This ratio is found by dividing the CO_2% into the CO% per volume.

Example: For an appliance in which the CO taken from a sample of flue gases was 25 ppm and the CO_2 was 5% the CO/CO_2 ratio would be:

$$25 \div 10\ 000 = 0.0025\% \text{ CO}$$

Therefore $\dfrac{0.0025\% \text{ CO concentration}}{5\% \ CO_2 \text{ concentration}} = 0.0005$

which suggests the appliance is burning safely.

Flue gas samples are usually taken by loosening the access plate at the top of the combustion chamber or by inserting a probe down below the draught diverter; samples taken *after* the draught diverter are suspect because the flue gases are likely to have been diluted by air being drawn in at the diverter skirt itself.

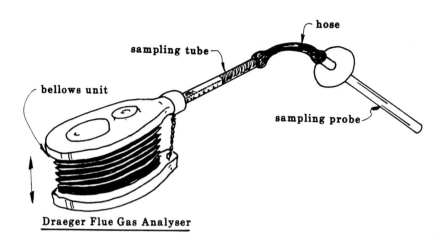

bellows unit

sampling tube

hose

sampling probe

Draeger Flue Gas Analyser

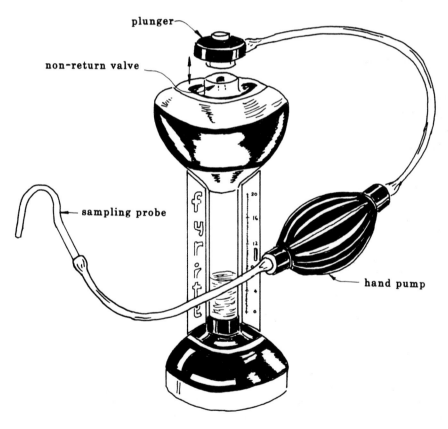

plunger

non-return valve

sampling probe

hand pump

Fyrite CO_2 Analyser

Combustion Analysis

Flue Efficiency in Gas Appliances

Relevant British Standard
BS 1756

Test for spillage

This test is carried out on open-flued appliances to ensure that no combustion products are entering the room via the draught diverter. The spillage test helps confirm the adequate provision of a suitable air supply and ensures a suitable updraught within the flue. The test is carried out as follows:

(1) Close all windows and doors to the room and light the appliance. Wait for between 5 and 10 min for the flue to warm up.
(2) Ignite a smoke match and hold it about 3 mm up inside the lower edge of the draught diverter, along its whole length, or perimeter. Spillage is indicated by the smoke discharging into the room.

Momentary spillage can be ignored, but if persistent the situation must be rectified. Sometimes, especially in cold weather conditions, one may need to wait longer than the initial 10 minutes to allow the flue to warm up sufficiently.

If an extractor fan is present in the same room as the appliance, or an adjacent room, the test should be carried out as above with any interconnecting door open and with the fan running. **Note**: extractor fans include those on cooker hoods and tumble driers, etc. If spillage occurs the fault will need to be rectified. Generally an extra 5000 mm^2 of free air ventilation will be sufficient in most cases to cure the problem.

Flue/appliance efficiency

The efficiency of the flue can be found by comparing the CO_2% with the flue gas temperature and adopting the following formula:

$$(0.343 \div CO_2\% + 0.009) \times (\text{flue gas temp} - \text{room temp})°C + 9.78 = \text{flue loss}$$

If one subtracts the energy due to flue loss from 100% one can see how efficiently an appliance is operating.

Example: Assume the CO_2 to be 5%, the flue gas temperature to be 180°C and the room temperature to be 21°C. Calculate the efficiency of the appliance. Thus:

$$(0.343 \div 5\% \ CO_2 + 0.009) \times (180°C - 21°C) + 9.78 = \% \text{ flue loss}$$

$$\text{Therefore } 0.0776 \times 159 + 9.78 = 22.1\% \text{ flue loss}$$

$$\text{Therefore } 100 - 22.1 = 77.9\% \text{ efficient}$$

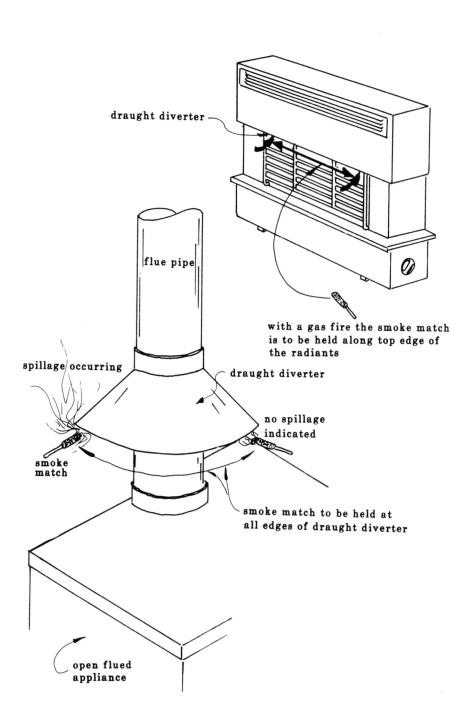

draught diverter

flue pipe

with a gas fire the smoke match
is to be held along top edge of
the radiants

spillage occurring

draught diverter

no spillage
indicated

smoke
match

smoke match to be held at
all edges of draught diverter

open flued
appliance

Flue Efficiency in Gas Appliances

Commissioning of Gas Installations

Upon the completion of any gas installation the system should be commissioned in stages to ensure its safe and efficient use. Most of the tests described below have been identified elsewhere in this book. Reference should be made to the appropriate pages if necessary. Some of the tests will not be applicable: e.g. spillage is not carried out on room-sealed appliances and should therefore be ignored.

Tests prior to commissioning

(1) Complete a general visual check of the system and flue routes for obvious defects;
(2) Carry out a soundness test (see page 162);
(3) Purge the gas carcass of air (see page 163);
(4) Carry out any pull tests to flue systems as necessary (see page 174);
(5) Identify the standing pressure at the meter (see page 164);
(6) Check the correct provision of a suitable air supply to the appliance (see page 178);
(7) Confirm the electrical connections are correct, i.e. earth continuity and polarity, and ensure the correct fuse has been fitted (see page 240);
(8) If applicable, confirm the boiler is filled with water.

Commissioning the appliance

(1) Check the standing pressure at the appliance (see page 164);
(2) Check the operation of the pilot flame and make adjustments if necessary; the flame's length should be about 25 mm and it should be playing on to the thermocouple;
(3) Check and adjust the working pressure of the appliance (see page 164);
(4) Identify and confirm the heat input to the appliance (see page 164);
(5) With the appliance alight, spray the final section of pipework to the burner from the inlet control valve with leak detection fluid to confirm its soundness;
(6) Check the operation of the flame failure device (e.g. by blowing out the pilot, etc.);
(7) Observe the general flame picture (a yellow flame indicates incomplete combustion of the fuel);
(8) Carry out a test for spillage (see page 184);
(9) Check the operation of the thermostat, including the high limit stat where applicable;
(10) Carry out any required flue gas analysis (see page 182).

Final system checks

Check the system working pressure to ensure the maximum pressure drop across the system does not exceed 1 mbar (see page 164).

Where a gas boiler is connected to a c.h. system, etc., the circulatory pipework will also need to be commissioned; in this connection the notes for commissioning c.h. systems should be sought (see page 146).

Handing over The working of the appliance should be demonstrated to the user and the best method of economic and efficient usage explained. All documentation supplied should be left with the owner/occupier who should be made aware of the benefits of a regular service contract to ensure that the equipment is maintained in an efficient and safe operating condition.

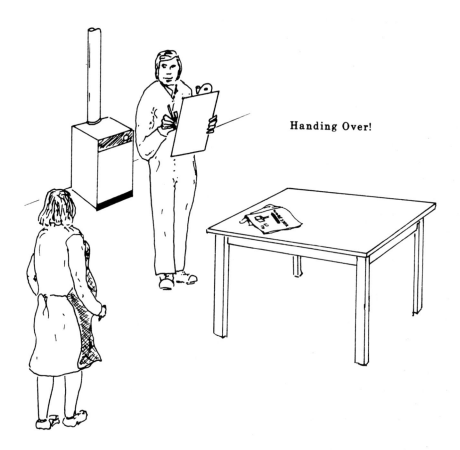

Handing Over!

Commissioning of Gas Installations

Maintenance and Servicing

When servicing a gas appliance such as a fire or boiler the procedure is identical to that followed when commissioning gas appliances except that, in addition, the appliance should be checked for carbon or soot build up. For example, the burners will need to be exposed and brushed or blown through with air to rid them of deposits. The heat exchanger should be brushed and vacuumed clean as necessary and the air intake checked for such things as cat fur and dust build up.

With open-flue appliances which connect to a chimney the appliance must be removed and the flue inspected; any debris from the collection space at the base of the chimney should be cleaned as necessary. When servicing it is always advisable to lay down a dust sheet; and a schedule should be completed to identify the tasks to be carried out. Should a boiler be connected to a c.h. system, this may form part of the service contract (see page 146).

When one goes to inspect, install, commission or service any gas appliance located within premises, an assumption is made that the gas service engineer last visiting the job is responsible for the whole installation. The fact that such an assumption is often made worries many people and the matter is open to debate as to its implications.

Provided the previous service engineer has made no adjustments to the gas rate (i.e. he has not adjusted any governing devices) and has not therefore changed the supply flow of gas to other appliances, there is little to worry about as to their operation, and by carrying out a visual inspection one would not be deemed to have validated the appliance's safe and efficient use. Where adjustment to the gas pressure at an appliance has been made, or an appliance installed, the rate of gas flow to other appliances may now be starved; in these circumstances a check on the system working pressure must be made (see page 164). If necessary a test nipple will need to be inserted at the furthest appliance.

The schedule shown opposite allows the installer or service engineer to complete a checklist which, upon completion, can be given to the owner/occupier; it informs them of the work completed and highlights any shortcomings.

Where an unsafe appliance has been located the appliance must be rectified or shut down and isolated; the schedule will highlight this. Where an urgent problem such as an unsafe appliance has been identified a copy of the schedule should be sent to the supplier of the gas for information and file. This will give the engineer some protection if the owner simply turns on the supply to the unsafe appliance as soon as the gas fitter leaves the premises. Schedules of this nature inform the client of possible faults and invariably give the installer more work, checking other appliances brought to the attention of the client.

Inspection & Testing to Gas Installation at:-

Address of Inspection: 52 MALDON ROAD, SPRING END, COLCHESTER ESSEX, COL 5FD

Date of Inspection: 11 OCT 93. Inspection/Work carried out by: J.CARTER

Gas Installer: J.CARTER
Address: 217 BROOK STREET, COLCHESTER ESSEX, CO2 17P
Telephone Nº: 0206 542.2.

Client: HARLER & BROWN ASSOCIATES
Address: 11 MARKET STREET, CHELMSFORD ESSEX, CM1 21P
Telephone Nº: 0245 9000

	Pass	Fail
20 mbar Soundness test on installation domestic installation (meter size 0.071 ft³/hr) max pressure drop = 4 mbar	✓	

Maximum Standing Pressure: 21 mbar
Maximum Working Pressure (maximum pressure loss permitted = 1 mbar): 0.8 mbar

Details of Servicing/Commissioning Carried Out

Appliances on Gas Supply

Item	Location/Appliance	Inspected & tested Yes	Inspected & tested No	Electrical Polarity Correct	Electrical Continuity Maintained	Fuse Rating	Standing pressure mbar	Working pressure mbar	Gas Rate Btu/hs	Gas Rate kW	Pilot Plane Correct & Adjusted	Flame failure device working correctly	Flame picture good	Thermostat operation correct	Ventilation Grill Required size	Ventilation Grill Pass	Ventilation Grill Fail	Flue test Pull Pass	Flue test Pull Fail	Flue test Spillage Pass	Flue test Spillage Fail	Flue Gas Analysis CO₂	Flue Gas Analysis CO₂/CO ratio	Temp °C	Efficiency%
1	KITCHEN - Boiler	✓		✓		3amp	21	17	66,000	19½	✓	✓	✓	✓	56cm²	✓		✓		✓		.04	00/80 5	0.006 260	79
2	KITCHEN - Cooker		✓																						
3	LOUNGE - fire		✓																						
4																									
5																									
6																									
7																									
8																									

In accordance with Regulation 34 of the Gas Safety (Installation & Use) Regulations 1984 I am duty bound by law to inform you that the gas appliances itemed...2 & 3... above have not been inspected by the gas installer herein and that their safe functioning cannot be guaranteed, therefore operation and safe usage rests with the owner/occupier of the said installation as signed on this schedule.

Gas Installers Signature: J.Carter   Owner/Occupier Signature: [signature]

Recommendations &/or Urgent Notification

THE OPEN FLUE ONLY EXTENDS 250mm UP ABOVE THE ROOF LINE. I RECOMMEND EXTENDING THIS HEIGHT TO 0.6m, THEREFORE BRINGING THE INSTALLATION UP TO CURRENT BRITISH STANDARD 5440 pt 1.

Maintenance and Servicing

Part
Oil Supplies

5

Properties and Combustion of Fuel Oils

Relevant British Standard
BS 5410

Liquid fuels in the United Kingdom are available in two grades: distillate and residual. Residual grades are very thick and are only used in heavy industry. Two types of distillate grade fuels are used for domestic burners namely:

☐ Class C2 28 second fuel or kerosine
☐ Class D 35 second fuel or gas oil.

Specification data for Class C2 and D distillate grade fuels

Property	Class C2	Class D
Colour	Clear	Red
Calorific value	46.3 kW/imp. gal	48.1 kW/imp. gal
Relative density	0.79	0.83
Viscosity (Redwood No 1 scale)	28 s	35 s
Solidification temperature	−40°C	−10°C
Flash point	38°C	55°C
Sulphur content	0.2	1.0

Note: Class D oil has a greater calorific value than Class C oil; thus Class D oil gives out more heat per volume and it is slightly cheaper in price. However, it is prone to solidification (waxing up) in cold weather and, because it has a high sulphur content, when combustion takes place sulphur dioxide (SO_2) is produced which, when combined with water, produces a weak sulphuric acid (H_2SO_4); this tends to corrode the heat exchanger more quickly, thus shortening the life of the boiler. As a result not many manufacturers recommend the use of Class D fuel and where it is used a preheater is often required to warm the fuel at the burner, thus reducing its viscosity.

The *viscosity* of a fuel governs its ability to flow easily. Treacle flows slowly whereas water flows very fast. The Redwood No 1 viscometer is a device which measures how long 50cm³ of a liquid, at 37°C (100°F), would take to flow through a hole of a defined size, the length of time being given in seconds.

Liquid fuel oils, like natural gas (or any other fuel), are known as hydrocarbons, being composed primarily of hydrogen and carbon atoms. The composition of both Class C and D oils is about 84% carbon and 16% hydrogen. Although the principles of the combustion process are the same as those for gas it will be found that where insufficient oxygen is available to consume the fuel, vast amounts of unburnt carbon (i.e. smoke) will be discharged from the flame; this unburnt carbon is deposited as soot in the heat exchanger and flue-way.

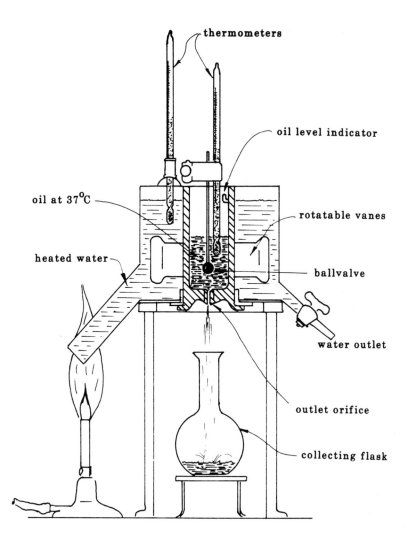

thermometers

oil level indicator

oil at 37°C

rotatable vanes

heated water

ballvalve

water outlet

outlet orifice

collecting flask

section through a Redwood N°1 Viscometer

Properties and Combustion of Fuel Oils

Oil Storage

Location of tanks

An oil storage tank should be sited in such a position that no danger would result from an oil spillage. The tank is ideally located externally; however, it may be located internally or below ground. Where the tank is closer to the building than 1.8 m a fire-resisting wall must be provided between the tank and the building itself.

Should the tank be located internally it must be sited within a fully enclosed tank chamber of non-combustible fire-resisting materials (ones which will offer fire resistance for no less than 1 hour). The base of the chamber should be of adequate structural strength and impervious to the passage of oil; thus, in the event of a major leak it will be capable of containing the full volume of oil. The base of the chamber should be laid to fall to an undrained sump, where a pump may be located to remove the oil. The entrance to the chamber should have a self-closing door and must be at high level above the catchpit area. Sufficient space must be provided in the chamber to allow adequate maintenance.

Tanks located below ground need to be anchored down to a concrete base and so constructed as to withstand any pressure imposed by the surrounding ground. Special treatment should be given to steel tanks so that they are able to resist corrosion.

Oil storage tanks above ground are usually mounted on supports and if these are of brick construction a damp-proof membrane will be required between metal tanks and the support to prevent corrosion. The minimum net capacity of the oil storage tank should be not less than 1250 litres; however, in the interests of economical deliveries it is advisable to have an oil tank with a capacity of at least 2700 litres (600 gal.). Tanks should be installed to slope down towards the drain cock, at a gradient of 20 mm/m.

The fill-up connection should be suitably capped and the vent fitted with a return bend to prevent the entry of water. The size of the vent should to be equal to the fill pipe and in no case smaller than 50 mm. The oil delivery tanker must be able to drive within 30 m of the tank, otherwise a large diameter pipe of 50–65 mm nominal bore (2–2½ in) will be required to extend the oil inlet to a suitable position. The oil outlet to the burner should be taken from the opposite end to the drain cock and slightly above the base, with an isolation valve fitted as necessary.

Oil level indicators Several designs of oil level indicator are available, including sight tubes and float gauges; or a dip stick could simply be inserted into the top of the vessel. The purpose of these devices is to indicate when a delivery of oil is required. Where a sight tube is used a valve should be located at its inlet to allow its removal for cleaning purposes and as a safeguard in case of breakage.

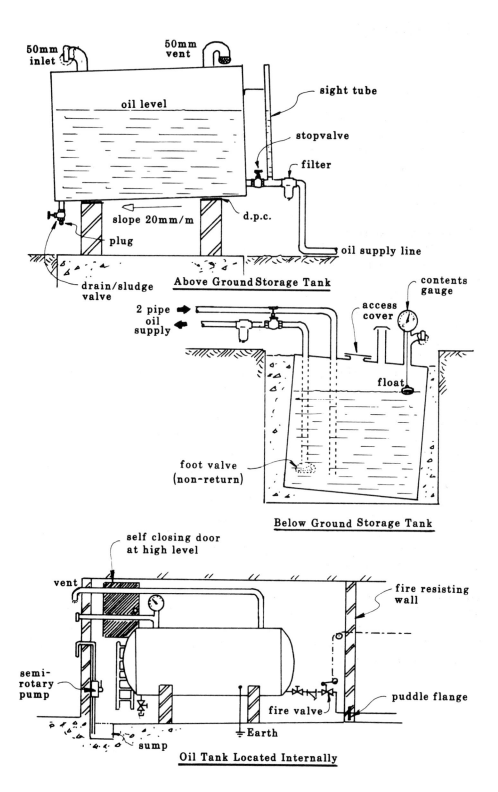

50mm
inlet

50mm
vent

oil level

sight tube

stopvalve

filter

slope 20mm/m

d.p.c.

plug

oil supply line

drain/sludge
valve

Above Ground Storage Tank

contents
gauge

access
cover

2 pipe
oil
supply

float

foot valve
(non-return)

Below Ground Storage Tank

self closing door
at high level

vent

fire resisting
wall

semi-
rotary
pump

puddle flange

fire valve

Earth

sump

Oil Tank Located Internally

Oil Storage

Oil Feed Pipework

Relevant British Standards
BS 799 and BS 5410

The pipe supplying oil to the burner should have a minimum diameter of 10 mm. It should incorporate two stopvalves, one at the storage tank and one by the appliance, and a filter and fire valve should also be located within its length. It is recommended that two filters be positioned for pressure jet burners, one at the tank and one prior to the burner. If just one filter is used this must be located at the boiler. The oil line should be run so as to avoid the trapping of air which may restrict or stop the flow of fuel. Where 35-second Class D fuel is used it may be necessary to insulate the pipework in exposed positions in order to prevent the fuel waxing up (solidifying) in winter.

Gravity feed systems Where the oil tank is fitted at a raised elevation above the burner its head pressure should not exceed 3 m to the underside of the tank; equally, to ensure a good flow this head should be at least 0.3 m. The actual size of the oil line will be dependent on the location and head pressure created (see table). Where high points cannot be avoided they should not be above the outlet of the tank and they should be fitted with a means of manual venting.

Maximum pipe length

Pipe dia	Head (m)			
(mm)	0.5	1.0	1.5	2.0
10	10	20	40	60
12	20	40	80	100

Sub-gravity system Where the bottom of the tank is less than 0.3 m above the burner it will be necessary to run a 10 mm return pipe (see figure). A spring-loaded non-return valve will be required on the supply pipe to prevent oil draining back to the tank. Alternatively, an oil deaerator, e.g. the Tiger-loop, may be used; this removes the air from the oil feed on a single pipe lift; the burner pump is piped to the Tiger-loop as shown, the latter being fitted close to the burner, although not inside the casing. Where the suction head is excessive an oil lifter can be used which draws the oil up to an elevated position from which it can flow by gravity to the burner.

The materials to be used for the oil supply line include fully annealed copper tube (grade Y), with manipulative type compression fittings being recommended (soft soldered joints should not be used, or low carbon steel pipe (not galvanised) may be chosen. Tapered threads should be cut on the pipe; running joints such as long screws, etc., should not be used.

Petroleum resisting compounds and PTFE tapes which remain slightly plastic make the most satisfactory joints; hemp and hard-setting jointing compounds should be avoided.

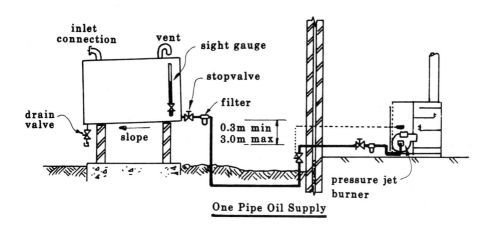

inlet connection
vent
sight gauge
stopvalve
filter
0.3m min
3.0m max
drain valve
slope
pressure jet burner

One Pipe Oil Supply

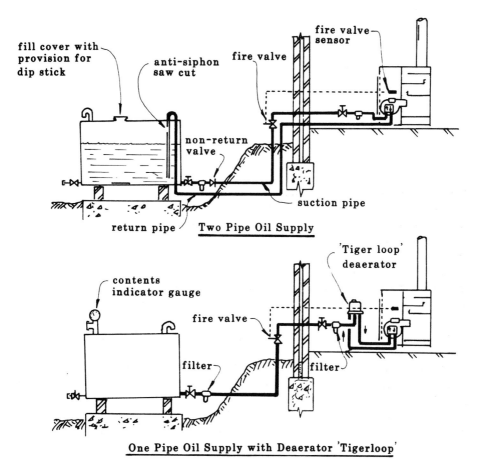

fill cover with provision for dip stick
anti-siphon saw cut
fire valve
fire valve sensor
non-return valve
suction pipe
return pipe

Two Pipe Oil Supply

'Tiger loop' deaerator
contents indicator gauge
fire valve
filter
filter

One Pipe Oil Supply with Deaerator 'Tigerloop'

Oil Feed Pipework

Controls Used on Oil Feed Pipework

Relevant British Standard
BS 5410

Filter This device is fitted in the pipeline from the oil tank to the burner and is designed to prevent the passage of unwanted particles that would otherwise block the burner jets of the appliance. Two types of filter will be found: those which contain a fine mesh, which can be removed for cleaning purposes; and those which contain a paper element, this being replaced periodically.

Fire valves These are special valves designed to shut off the flow of oil to an oil-burning appliance automatically in the case of a fire. There are three basic types: the *fusible link* type, the *leaded handle* type and the *pressure* type. With the fusible link type the valve is held open by a tensioned wire and the arrangement is set up as shown in the figure.

The fusible link has a low melting temperature and in the event of a fire will break; the valve will then close, assisted by a weight or attached spring. The fusible link type would normally only be used in a purpose-made boiler house.

For domestic oil-fired installations either the leaded handle or pressure type is generally used. The leaded handle type is probably the most common and consists of a small spring-loaded stopvalve in which the thread in the wheel-head is made of a low melting solder. In the event of a fire the solder melts and the valve springs shut. The pressure type consists of a bellows type valve which is connected to a heat-sensitive bulb. This bulb overheats in the event of a fire and causes an increased vapour pressure in the bellows, resulting in its expansion; thus the valve is closed.

The heat sensitive element should be located above the combustion chamber at a minimum height of 1 m, although where this is impracticable the element can be fitted as close as possible to the appliance, at a similar level; however, it must be in the same room as the appliance.

When an external tank is being used the fire valve should be fitted as close as possible to the point where the supply enters the building, ideally externally.

Constant oil level control This device will only be found on the supply to a vaporising burner. It is designed to control the feed of oil to the appliance at a constant pressure and flow, irrespective of the oil level in the storage tank. Should the oil flow directly to the burner from the oil tank it would flood the appliance, as its operation is dependent on a maintained tray of oil at its base. The vapour from the oil rises and is ignited or, in the case of a wallflame burner, drawn up to be dispersed round the perimeter of the combustion chamber.

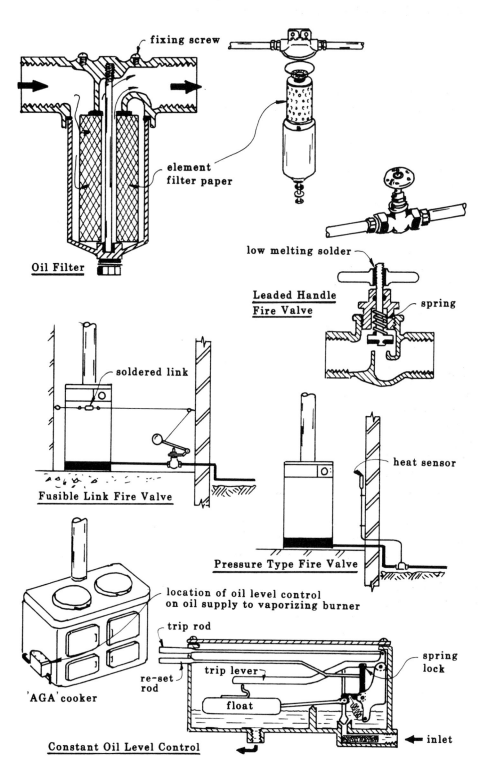

fixing screw

element
filter paper

Oil Filter

low melting solder

Leaded Handle
Fire Valve

spring

soldered link

Fusible Link Fire Valve

heat sensor

Pressure Type Fire Valve

location of oil level control
on oil supply to vaporizing burner

trip rod

'AGA' cooker

re-set
rod

trip lever

float

spring
lock

inlet

Constant Oil Level Control

Controls Used on Oil Feed Pipework

Pressure Jet Burners 1

Relevant British Standards
BS 799 and BS 5410

All oil boilers sold today are of this design and consist of a self-contained package. The burner is reasonably quiet in operation and basically works as follows:

(1) Should the thermostat require heat the motor runs, rotating a fan which allows air to be blown into the combustion chamber. This rids the appliance of any residual vapour.
(2) After a set time a solenoid opens to allow oil to be pumped through a small nozzle into the combustion chamber in a fine spray (a process referred to as atomizing), where it mixes with the air. Simultaneously a spark is generated at two electrodes inside the chamber, located at the tip of the nozzle to allow ignition to occur.
(3) When the flame is detected by a photoelectric cell, which distinguishes light rays from within the combustion chamber, the electrodes cease to spark.
(4) The motor continues to run, allowing fuel and air into the combustion chamber until the thermostat is satisfied, at which the solenoid closes, allowing the motor to stop.

The burner unit is located in a steel heat exchanger, which consists of a rectangular combustion chamber. The waterway surrounds the chamber with flow-and-return connections at high and low level. Oil burners have a large void through which the hot flue gases pass to prevent the build-up of soot; therefore a series of baffle plates are located within the chamber to direct the flow of hot gases on to the walls of the heat exchanger.

Upon initial commissioning it is essential to check that these plates are in position in order to avoid damage to the unit. The hot gases pass up through the chamber and are expelled from either an open or balanced flue arrangement.

When installing burners with a two-pipe oil supply or a one-pipe system incorporating a Tiger-loop, it is essential to install a by-pass screw into the pump, as supplied by the manufacturer, thus preventing its damage.

On first lighting up a new burner which has also had a new supply run it may take several attempts to ignite. It is best to bleed the supply pipe of air by allowing the oil to discharge into a container. A combined air bleed manifold and pressure gauge (to register 0–20 bar) should be connected to the appropriate oil pump connection on the burner and the thermostat turned up (assuming the electricity supply is on and calling for heat).

When the motor starts it may be necessary, on the one pipe system, to open the bleed screw on the test manifold to remove the air. If the burner locks out, wait about a minute or so and reset the lock-out button on the control box. Once the burner fires up, the burner pressure should be checked and adjusted to the correct pressure, as recommended by the manufacturer, usually in the region of 7 bar (100 lb/in^2).

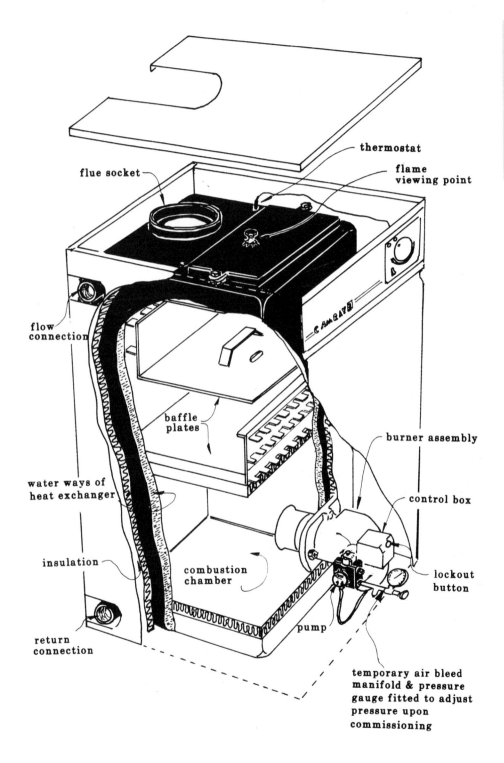

thermostat

flame
viewing point

flue socket

flow
connection

baffle
plates

burner assembly

water ways of
heat exchanger

control box

insulation

combustion
chamber

lockout
button

return
connection

pump

temporary air bleed
manifold & pressure
gauge fitted to adjust
pressure upon
commissioning

Pressure Jet Burners 1

Pressure Jet Burners 2

The pressure jet burner consists of the following component parts:

Control box An electrical device which controls the operation of the burner in a pre-determined sequence. In the event of a boiler malfunction the control box will cause the sequence of operations to stop and dislodge a trip switch (referred to as lock-out) causing the complete shutdown of the appliance. To restart the appliance a reset button needs to be manually depressed. **Note**: some burners will automatically carry out a second attempt at ignition.

Electric motor Provided to rotate the air fan and fuel pump simultaneously.

Air fan Designed to supply air, under pressure, into the combustion chamber. The amount of air supplied will be dependent upon the air shutter opening which is adjusted during commissioning.

Fuel pump A device which receives the oil from the oil storage tank at atmospheric pressure and increases its pressure to around 7 bar (approximately 100 lb/in^2) thus forcing the oil through the small hole in the nozzle.

Solenoid valve (a magnetic valve) This device prevents oil entering the combustion chamber until the appliance has been purged with air. It also allows the oil to come up to pressure, as required, prior to entry into the chamber, allowing the fine spray to be formed. When the appliance has reached temperature the solenoid also gives a quick shutdown of the fuel.

Air diffuser A device which causes the air to turbulate and swirl around, allowing for a better fuel/air mixture.

Electrodes Two conductors which are critically positioned at the nozzle tip to allow a flow of electricity to jump between their ends; this causes a spark, which is used to ignite the fuel.

Transformer The device which increases the voltage of the power supply to some 10 000 V, which will be sufficient to produce the required spark at the electrode tips.

High tension (HT) leads The cable which carries the high voltage to the electrodes, similar to spark plug leads on a car. **Note**: some burners, such as the Riello, do not use these leads; the electrodes simply plug into the transformer located in the control box.

Photocell A device which detects light. It consists of a small resistor which, when exposed to light rays, makes or breaks an electrical circuit.

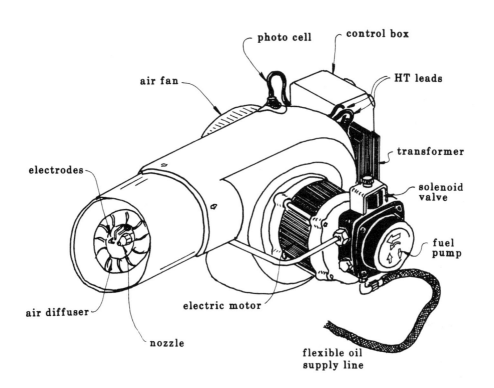

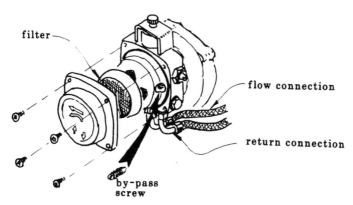

Inserting a by-pass screw into the pump
for use with a two pipe system

Pressure Jet Burners 2

Pressure Jet Burners 3

Nozzle and ignition setting

The nozzle is the device which atomises the fuel and causes it to leave the burner in a swirling motion in readiness for ignition by the electrodes. Stamped on the body of the nozzle will be the spray angle in degrees and the flow rate in US gallons, plus an indication of the type of spray formation. This information must always be checked, i.e. upon initial set up and servicing to ensure compliance with the manufacturer's data and efficient and economic use. The spray angle for domestic boilers is usually 45°, 60° or 80°, the angle of spray being governed by the length of the combustion chamber. Generally a long chamber requires a small spray angle.

The flow rate is the volume of oil in 1 h (the US gallon is used because the pressure jet burner was developed in the United States). Where data are unavailable the following formula can be used to give a guide to a suggested flow rate.

$$\text{Boiler output} \div 0.8 \div \text{calorific value of oil} \times 1.2 = \text{US gallons/h}$$

Note: 0.8 suggests the boiler is 80% efficient and the 1.2 changes imperial gallons to US gallons.

Example: Identify a suggested nozzle size for a pressure jet burner when the output is 17½ kW and the fuel used is Class C2, 28-second oil. (**Note**: the CV for Class C2 fuel is 46.3 kW/gal.)

Answer: $17.5 \div 0.8 \div 46.3 \times 1.2 = 0.57$ US gal./h
for which the nearest nozzle flow size is 0.6 US gal./h

The electrode setting needs to be checked. Where a problem is encountered with ignition bear in mind that the ignition spark will take the shortest distance when jumping a gap; so ensure that the narrowest gap is between the electrode tips and not between one electrode and, say, the nozzle. Where possible, set the electrodes alongside the flat of a nozzle rather than its corner, thus giving a bigger gap clearance.

Also check for carbon build-up on the electrode tips. As a general rule the electrodes are positioned 3 mm apart and 15 mm away from the centre line of the nozzle outlet, usually slightly forward of the nozzle itself.

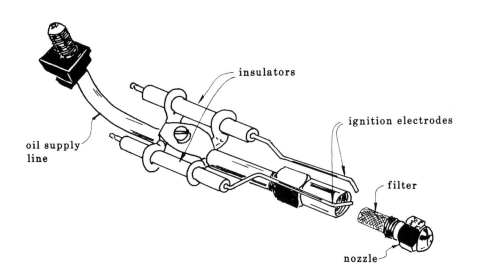

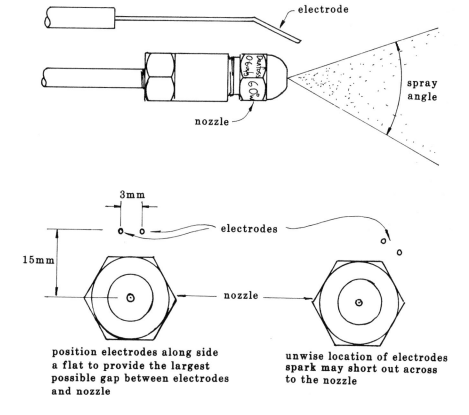

position electrodes along side
a flat to provide the largest
possible gap between electrodes
and nozzle

unwise location of electrodes
spark may short out across
to the nozzle

Pressure Jet Burners 3

Open Flued Appliances

Where a traditional brick chimney is encountered it should be suitably lined with a stainless steel liner. It may be necessary to install loose-fill insulation material in the void between the liner and the exposed chimney in order to avoid condensation problems. Any liner or flue pipe used should be of the same diameter as the outlet from the appliance. Chimneys and flue pipes should be as straight as possible and be free from bends exceeding 135° (45°).

The terminal location for an open-flued system should be above any excessive positive or negative pressure zones: generally a distance of 1 m above the roof line is sufficient. The height of the flue should not exceed 6 m in length, as this may cause excessive flue draught problems. The flue draught for pressure jet burners should be around 0.09 mbar (0.035 in water gauge (wg)). However, if the draught exceeds 0.4 mbar (0.15 in wg) a draught stabiliser should be fitted.

The *draught stabiliser* consists of a hinged flap, usually fitted to the base of a chimney. It is fitted in the same room as the appliance and its purpose is to ensure that the chimney maintains the required draught condition. Should the draught within the chimney be too high, because of high winds, for example, the flap will open, permitting air to enter directly into the chimney and thus reducing the draught through the boiler. The draught stabiliser is also forced open should the pressure inside the chimney become excessive.

Air supply requirements

Open flued oil burning appliances with an input below 5 kW do not require any additional air requirements. Where the input is in excess of 5 kW the room in which the appliance is installed must have an air grille with a minimum effective free area of 550 mm^2 for every kW in excess of 5 kW.

Example: A 22 kW open flued oil burner requires an effective grille size of:

$$22 - 5 = 17; \text{ thus } 17 \times 550 = 9350 \text{ mm}^2$$

Where a draught stabiliser is fitted this grille size should be doubled.

Should a boiler be located in a cupboard or small confined space, suitable air grilles should be provided for cooling purposes. These would consist of grilles at high and low level to the outside air of 550 mm^2 per kW input; this figure is doubled where the grille connects to an internal room, which itself needs to be suitably vented. The effective free area of a grille is defined under ventilation for gas appliances (see page 178). Where the flue gas temperature does not exceed 260°C the flue pipe or brick chimney, etc., should be installed in the same way as for gas installations. However, if this temperature is in excess of 260°C, Part J of the Building Regulations should be sought and installation of pipes, etc., should be as for solid fuel appliances.

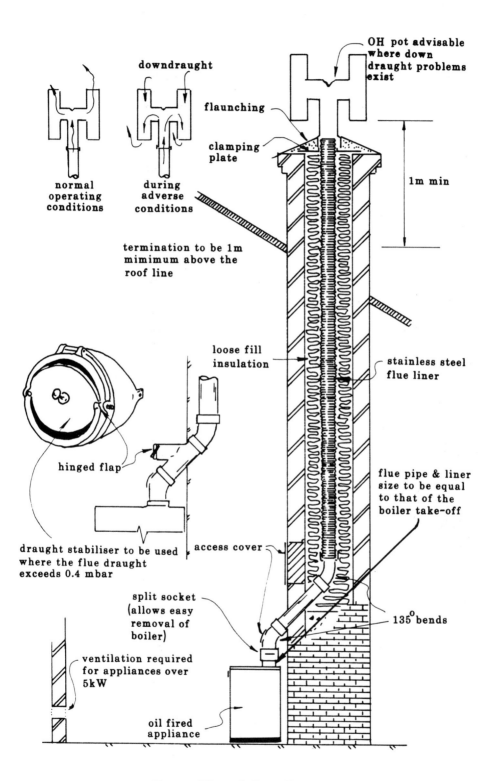

downdraught

OH pot advisable where down draught problems exist

flaunching

clamping plate

1m min

normal operating conditions

during adverse conditions

termination to be 1m mimimum above the roof line

loose fill insulation

stainless steel flue liner

hinged flap

flue pipe & liner size to be equal to that of the boiler take-off

draught stabiliser to be used where the flue draught exceeds 0.4 mbar

access cover

split socket (allows easy removal of boiler)

135° bends

ventilation required for appliances over 5kW

oil fired appliance

Open Flued Appliances

Room Sealed Appliances

Relevant British Standard
BS 5410

With the forced draught of air into the combustion chamber in the pressure jet burner the extraction of flue gases is that little bit easier and as a result the installation of the balanced flued appliance is often favoured. This allows for a faster installation without the need to worry about flue draughts and permanent ventilation, as required with open flued systems. The air required for combustion purposes is drawn directly from the outside environment.

Where a balanced flued boiler is to be installed it is essential that the casing is securely sealed; otherwise, should the pressure outside the building be in excess of that within the room, incomplete combustion will occur, resulting in smoke and carbon emissions from the flue terminal. This will generally stain the brickwork. The principle of the balanced flue relies on the pressure within the boiler casing and combustion chamber to be the same as that at the point of flue gas discharge.

Note that some boilers need to have an air blanking plate inserted in the base to seal the unit, as the boiler manufacturer often uses the same casings for both open and balanced flued appliances.

Boilers operating with Class D 35 s gas oil are not permitted to have their products discharged at low level. Where the terminal is located it is essential that a terminal guard be fitted for the protection of any person who might otherwise come into contact with the terminal itself. The location of terminal positions should be as indicated in the following table:

Suitable room-sealed terminal locations for oil appliances

Terminal position	Minimum distance (mm)
From a surface facing a terminal	2000
Vertically from a terminal on the same wall	1500
Horizontally from a terminal on the same wall	1200
From an internal corner	900
From an external corner	600
Horizontally from any opening, e.g. window or door	600
Directly below any opening, e.g. window or door	600
Below the eaves or discharge pipework	600

It is possible to have a room-sealed arrangement with a vertical balanced flue for use where it may not be possible to discharge through an external wall. Where a vertical balanced flue is used it must terminate at least 600 mm from any wall face.

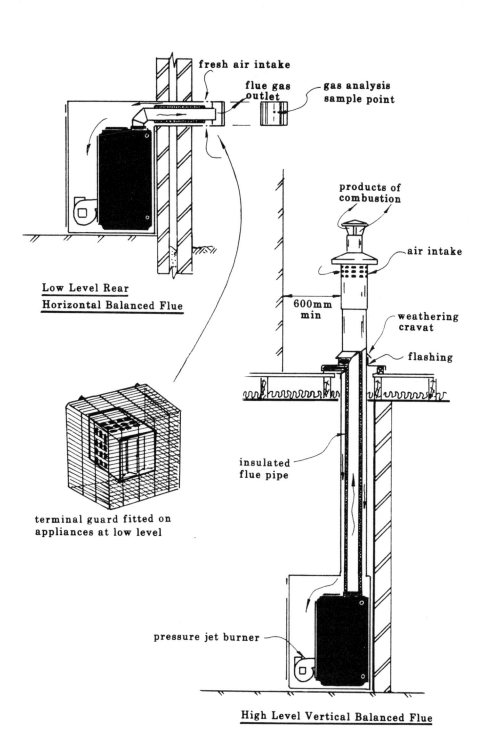

fresh air intake

flue gas outlet

gas analysis sample point

Low Level Rear Horizontal Balanced Flue

products of combustion

air intake

600mm min

weathering cravat

flashing

insulated flue pipe

terminal guard fitted on appliances at low level

pressure jet burner

High Level Vertical Balanced Flue

Room Sealed Appliances

Vaporising Burners

Relevant British Standards
BS 799 and BS 5410

The vaporising burner is no longer installed to serve as a boiler, its use being generally restricted to AGA type cookers. The burner works on the principle of warming the oil to cause it to change into its gaseous state where it can readily mix with the oxygen in the air which is required to support combustion; in this form it can be ignited easily. There are three designs of vaporising burner:

☐ The natural draught vaporising burner (natural draught pot burner);
☐ The forced draught vaporising burner (fan assisted pot burner);
☐ The wallflame or rotary vaporising burner.

To ensure that not too much oil is fed to the burner at any time and at any pressure, a constant oil level control is fitted at the base of the appliance. This contains a float which allows a small quantity of oil into its reservoir (see page 198).

The natural draught 'pot' burner is the simplest type of oil burner and consists of a circular container (the pot) with a series of holes through which air can pass to the combustion chamber; basically it works as follows:

(1) When the appliance requires heat the igniter is activated, which glows red hot; then a solenoid valve opens which allows oil to trickle into the burner from a constant oil level control. Note that some burners are ignited manually with a lighted match or taper;
(2) The oil spreads as a thin film over the base of the burner and the vapour from the oil is ignited (it is the vapour which burns, not the liquid);
3) As the fuel burns it generates heat which speeds up the vaporising process and thus the flame burns more fiercely and rises up the chamber to obtain sufficient air for combustion;
(4) Eventually the flame burns out at the top of the pot where the waterways of a heat exchanger may be located;
(5) When the thermostat is satisfied the fuel is cut off to the burner. Note that the appliance will continue to burn owing to the oil and vapour in the pot.

It is most important that no attempt should be made to relight a warm or hot burner as the ignition of the fuel vapour could blow back with explosive force. The forced draught burner is identical except that air is blown into the combustion chamber.

In the wallflame burner the oil is drawn up from a well at the base of the boiler through two oil distribution tubes, which spin in the centre of the appliance when heat is required. The oil is thrown out to the rim of the combustion chamber and is ignited by a sparking high tension ignition electrode. As the droplets fall they are quickly vaporised by the previously heated rim heater and rise to mix with the oxygen, also blown through the centre. The flame quickly becomes established and burns up around the perimeter of the chamber in a wall of fire.

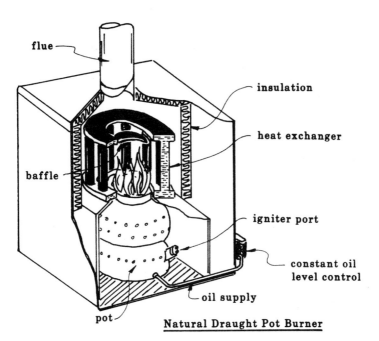

flue

insulation

heat exchanger

baffle

igniter port

constant oil
level control

oil supply

pot

Natural Draught Pot Burner

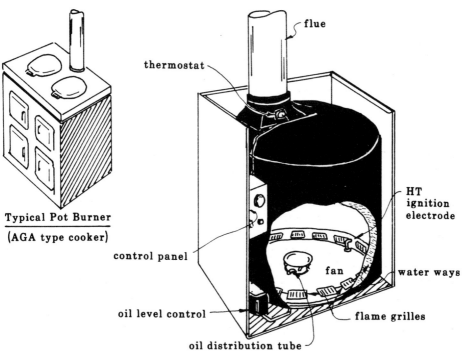

Typical Pot Burner

(AGA type cooker)

thermostat

flue

HT
ignition
electrode

control panel

fan

water ways

oil level control

flame grilles

oil distribution tube

Wallflame Burner

Vaporising Burners

Combustion Efficiency Testing

Relevant British Standard
BS 1756

When commissioning an oil burning appliance it is important to ensure its maximum working efficiency and that no undue smoke is produced. This can be achieved by taking the following measurements in the section of flue directly above the appliance, ensuring that any hole drilled is plugged on completion. In the case of the balanced flue appliance measurements should be taken at the terminal location. The equipment used to administer these tests may be electronic or more traditional as described below. In all cases allow the appliance an initial 10–15 minute warm-up period.

Flue draught Not applicable to balanced flued appliances. A probe is inserted into the flue to give a draught reading which should be as recommended by the manufacturer. A typical draught is 0.09 mbar (0.35 in wg).

Smoke reading A test in which a special pump is used to draw a quantity of flue gas through a special filter paper. A clean dry piece of paper is inserted into the pump and the probe inserted into the test hole. Ten suctions are given to the pump handle and the paper inspected for signs of discoloration. Where any signs are indicated the sample is held against a smoke gauge supplied. A reading of 0–1 should be achieved. To reduce the smoke content the air supply will need to be increased to the combustion process by adjusting the air shutter on the burner.

CO_2 percentage This test is carried out in order to find the carbon dioxide content within the flue gas, and it is used in conjunction with the flue temperature to find the efficiency of the appliance. Once the smoke has been adjusted to allow the minimum amount of air possible the CO_2 reading can be taken. Ideally one is looking for as high a $CO_2\%$ as possible, thus indicating greater theoretical combustion. Closing the air shutter raises the CO_2 but increases the smoke emissions. If the air shutter is altered at this stage the smoke reading must be retaken. The CO_2 indicator consists of a special container filled with a liquid capable of absorbing CO_2.

When the probe is inserted into the test hole the hand pump is operated to purge the connecting hose. The plunger (see figure) is now connected to the container and held down, causing a non-return valve in the CO_2 indicator to open. The pump is depressed 18 times and during the 18th deflation the plunger is removed. The container is now inverted several times; this causes the CO_2 to be absorbed into the liquid. With the gas absorbed into the liquid a partial vacuum is formed in the container and a diaphragm at the base of the vessel flexes upwards, which raises the level of the fluid. The $CO_2\%$ can now simply be read from a scale at the side of the container.

Flue gas temperature This is simply taken, using a flue gas thermometer or electronic pyrometer probe inserted into the test hole.

Flue/appliance efficiency This is found by comparing the CO_2% against the flue gas temperature, usually achieved using a sliding chart or by calculation using the following formula:

$$(0.477 \div CO_2\% + 0.0072) \times (\text{flue gas temp} - \text{room temp})°C + 6.2 = \% \text{ flue loss}$$

Thus, if you subtract the energy lost up the flue from 100% an indication of the efficiency of the appliance will be seen.

Example: With 11% CO_2, a flue gas temperature of 195°C and a room temperature of 21°C the appliance will be operating at:

$$(0.477 \div 11\% + 0.0072) \times (195 - 21)°C + 6.2 = \% \text{ flue loss}$$

$$\text{therefore} \quad 0.0506 \times 174 + 6.2 = 15\% \text{ flue loss}$$

$$\text{therefore} \quad 100 - 15 = 75\% \text{ efficient}$$

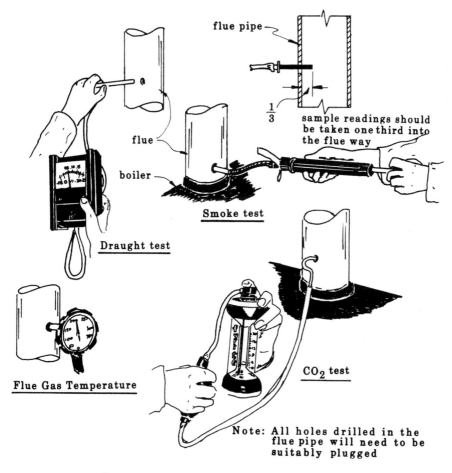

flue pipe

$\frac{1}{3}$

sample readings should be taken one third into the flue way

flue pipe

flue

boiler

Smoke test

Draught test

Flue Gas Temperature

CO_2 test

Note: All holes drilled in the flue pipe will need to be suitably plugged

Combustion Efficiency Testing

Commissioning and Fault Diagnosis

Relevant British Standard
BS 5410

When an appliance has been installed it should be commissioned in accordance with the manufacturer's instructions. However, the following checklist may be considered to avoid lengthy fault-finding problems in setting the burners up in such a way as to ensure their maximum efficiency.

Tests prior to ignition of the appliance

(1) Complete a general visual check of the system and flue to identify obvious defects.
(2) Disconnect the oil supply as close as possible to the burner, bleed the supply pipe and drain 4–5 litres of oil into a clean container. Inspect the oil for water or any impurities, repeat as necessary. Do not reconnect until clear, as damage may occur to the unit. Clean all filters and de-sludge the oil tank if necessary.
(3) Look for leaks in the oil supply pipework, inspecting the tank also.
(4) Check that the baffles are correctly positioned inside the boiler combustion chamber.
(5) Drill a hole in the flue-way to serve as a test point for open flue appliances, where applicable.
(6a) With *balanced flue appliances*, is the casing sealed correctly from the room?
(6b) With *open flue appliances*, is there an adequate supply of air for combustion?
(7) Confirm the electrical connections are correct, i.e. earth continuity and polarity, and ensure that the correct fuse is fitted (see page 240).
(8) Check the system is filled with water.
(9) Remove the burner as necessary to check:
  (a) the by-pass screw is fitted where applicable (see page 203)
  (b) the correct nozzle is fitted (see manufacturer's instructions)
  (c) the electrode settings are correct (see manufacturer's instructions)
  Fit the combined bleed manifold and replace the burner.

Commissioning the appliance

(1) Ignite the appliance and adjust the burner pressure (see page 200);
(2) Check the lock-out function; either cover the photoelectric cell or remove the solenoid coil to simulate flame failure;
(3) Check the operation of any thermostats fitted; where a high limit stat is fitted remove the control thermostat phial to simulate its malfunction;
(4) Carry out combustion efficiency tests (see page 212);
(5) Check the manual operation of the fire valve to ensure its operation;
(6) Remove the bleed manifold and give a final check for leaks.

Fault finding: pressure jet burners

Symptoms	Possible causes
Burner will not run although no lockout light has come on	No power is reaching the burner (i) Power is off (ii) Timeclock/programmer is in the *off* position (iii) Fuse has blown (iv) Thermostat not calling for heat
Burner will not run and the lockout light is indicated	Motor not running (i) Motor defective (ii) Motor capacitor defective (iii) Fuel pump seized To confirm (iii) remove pump and restart burner; if motor runs pump is at fault.
Burner runs but goes into lockout without firing	Starvation of fuel at the burner nozzle (i) Fuel tank empty (ii) Valves on fuel line shut (iii) Fuel pump drive sheared or worn; repair or renew Nozzle not atomizing fuel – renew nozzle Spark not reaching fuel (i) Electrodes incorrectly set (ii) Electrodes earthing out (iii) Defective high tension leads (iv) Ignition transformer defective
Burner runs and fires but goes into lockout	Dirty or defective photocell Starvation of fuel (i) Partially blocked filter. **Note**: in winter the fuel may be waxing up (ii) Valves in fuel line shut (iii) Air in fuel, e.g. oil level low (iv) Water in fuel (v) Faulty solenoid valve Faulty control box Blockage in flue, possibly causing back pressure in the combustion chamber
Burner runs and fires but fails to switch off	Faulty thermostat (i) Defective (ii) Loose connections (iii) Not correctly housed in appliance

Part 6

Electrical Work

Basic Electrical Theory

Electricity is the flow of free electrons moving along a conductor. Free electrons are the infinitely small electrical charges moving round an atom. Millions of atoms make up a solid conductor, such as copper, but the electrical charge is not felt because all the electrons are moving in all directions. It is when all these electrons are made to travel in the same direction that the electrical *current* is produced. A flow of about 6 billion electrons per second produces a current of about 1 *ampere*.

Electricity will not flow through certain materials such as wood or plastic because there are no free electrons in these materials and they are referred to as insulators. To make the free electrons flow in the same direction an electromotive force (emf), expressed in *volts*, needs to be applied, such as that produced by a battery or generator. The flow of free electrons can only occur if the electrons have somewhere to go, which is achieved by allowing the electrons to flow in a circuit. Break the circuit and the electron flow (current) will stop.

Resistance As the electrons flow through the cable they slowly lose their ability to move (emf). When the electrons are forced to pass through thin sections they all rub together in order to get through; thus they are restricted and once again there is a drop in emf. This voltage drop is due to the resistance of the system and its components; resistance is measured in *ohms* (Ω).

The relationship between current (amps), emf (volts) and resistance (ohms) is expressed in *Ohm's Law*:

volts ÷ amps = ohms; similarly volts ÷ ohms = amps and amps × ohms = volts

Example: where a 12 volt battery is used to serve a circuit in which the total resistance is 4 ohms the current flow will be 12 ÷ 4 = 3 amps. **Note:** increasing the voltage increases the amperes.

Power When the electrical energy produced is used up very quickly, such as in passing through a light bulb filament, heat energy is produced, due to resistance; this is referred to as power and is measured in *watts*. Again there is a relationship between current and the emf:

volts × amps = watts, watts ÷ volts = amps and watts ÷ amps = volts

So in the previous example, where the 12-volt battery had a current flow of 3 amps, the power would have been 12 × 3 = 36 watts.

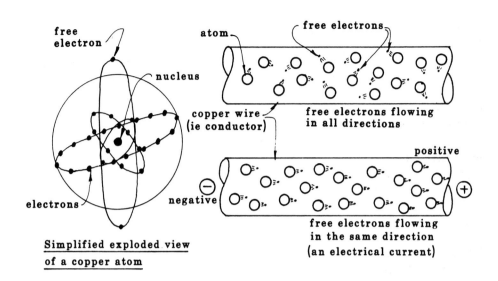

free electron

nucleus

electrons

Simplified exploded view of a copper atom

atom

free electrons

copper wire (ie conductor)

free electrons flowing in all directions

positive

negative

free electrons flowing in the same direction (an electrical current)

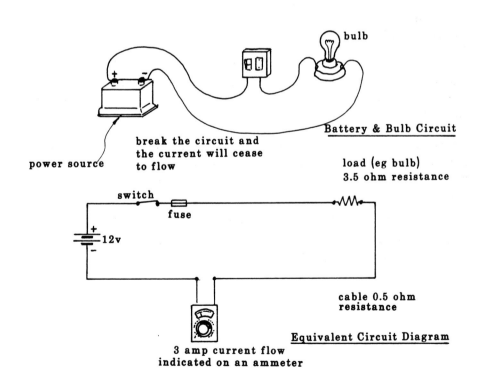

bulb

Battery & Bulb Circuit

power source

break the circuit and the current will cease to flow

load (eg bulb) 3.5 ohm resistance

switch

fuse

12v

cable 0.5 ohm resistance

Equivalent Circuit Diagram

3 amp current flow indicated on an ammeter

Basic Electrical Theory

Electrical Current

Simple circuits

Series circuit A system where the current flow is made to pass through each resistor (e.g. a bulb) in the circuit. Because of this the resistance slowly diminishes as the current flows round the system and as a result the total voltage of all the bulbs must not exceed the total available voltage, otherwise the bulbs will not glow sufficiently.

Parallel circuit In this system the same emf is applied to each element and as a result the voltage at all the bulbs needs to be equal to that of the battery. The bulbs will glow just as brightly whether one is removed or not, but the life of the battery is simply reduced as more bulbs are used, i.e. more power is used. With this system when a bulb blows the others stay alight whereas with a system in series when one bulb blows it breaks the circuit and all the bulbs go out. The lighting in a house is wired up in parallel and Christmas tree lights would be a typical example of a series circuit.

Direct current (d.c.) In a d.c. electrical circuit the electron flow is in the same direction all the time. One example would be a battery in which a metal such as lead slowly destroys another metal such as zinc (see page 26, Corrosion) and in so doing creates a small emf; d.c. is also produced when a thermocouple is heated.

Alternating current (a.c.) With a.c. the electron flow travels continually back and forth. This is a result of the way in which the current is produced. A.C. is produced by moving a magnet in and out or around a coil of wire, which is wound on to a soft iron core. The movement of the magnetic field around the wire causes electrons to flow, providing the circuit is made. As long as the magnet is moving the current will flow.

    The electric generator at the power station produces a.c. on this principle: one wire is connected to the overhead power cable, the other to 'earth'. Thus, when the cable in the home is connected to earth a circuit is formed and the current flows. From this we can deduce that in our home, which is supplied with 240 V a.c., the current flowing to an appliance is flowing from the phase (live) wire and then from the neutral wire (which is, in effect, connected to earth); the current flows continuously back and forth at a rate of 50 times a second (50 Hertz).

Three phase supply An a.c. supply where not only one circuit/wire is connected to the generator but in fact three. This gives three lots of current in sequence (three phases). In industry, where large motors are used, a single phase supply of 240 V would not produce sufficient power to operate the plant. The three phases, when all circuits are wired to the machinery, give an approximate emf of 415 V.

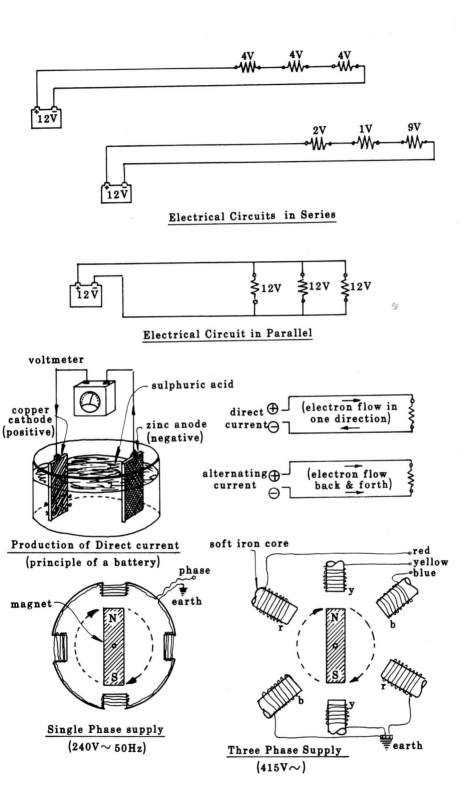

Electrical Circuits in Series

Electrical Circuit in Parallel

voltmeter

copper cathode (positive)

sulphuric acid

zinc anode (negative)

direct current $\oplus$ $\ominus$ (electron flow in one direction)

alternating current $\oplus$ $\ominus$ (electron flow back & forth)

Production of Direct current
(principle of a battery)

soft iron core

red
yellow
blue

magnet

phase

earth

Single Phase supply
(240V ∼ 50Hz)

Three Phase Supply
(415V∼)

earth

Electrical Current

Electrical Supply

Relevant British Standard
BS 7671

Electrical supply to the home

The electrical supply to the domestic dwelling is distributed by a network of above- and below-ground cables, via various substations; these run to the home in one of the following three distinct systems:

☐ *TT system*: Designed for older properties with overhead power supply; the earth is achieved by an earth stake.
☐ *TN-S system*: An older system which brings in the supply from below ground level. A metallic armoured casing encloses the phase and neutral cables. The earth is made by connection to the metal casing.
☐ *TN-C-S system*: Also called the PME system. The system most commonly used today; it utilises a special concentric cable, the phase being surrounded by the neutral. The earth and neutral are the same outer wire of the cable; it is only at the house that the earth is separated from the neutral. **Remember**: neutral and earth are to all intents and purposes the same thing.

The conductors terminate in an electricity authority sealed fuse unit at the dwelling. From here the earthing cable is run to the main terminal block; the phase and neutral are fed through a meter and into the main consumer unit where a double pole switch is incorporated to isolate the system. A fuse or circuit-breaker is incorporated to provide automatic protection to the system. Three wires are run from the incoming supply to the various circuits, these being:

☐ *The phase conductor* commonly called the live wire, although one should note that when the circuit is made the neutral also becomes live!
☐ *The neutral conductor*.
☐ *The circuit protective conductor* (cpc), commonly called the earth.

The earth conductor is designed as a back-up. Should a fault develop within the system where a live cable touches some metallic part (e.g. the appliance itself) the electric current flows through the cpc and travels down to earth. If you were to touch the metal part you would be prevented from getting a shock because your body would offer a greater resistance to the electron flow than the cpc. All the metalwork in an installation must ultimately be connected to the main earth terminal block.

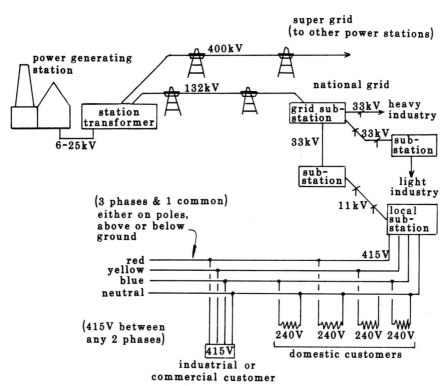

super grid
(to other power stations)

power generating
station

400kV

132kV

national grid

station
transformer

grid sub-
station

33kV heavy
industry

6-25kV

33kV

33kV sub-
station

sub-
station

light
industry

11kV

local
sub-
station

(3 phases & 1 common)
either on poles,
above or below
ground

415V

red
yellow
blue
neutral

(415V between
any 2 phases)

240V 240V 240V 240V

domestic customers

415V

industrial or
commercial customer

Single & Three Phase
Connections from generator
to Consumer

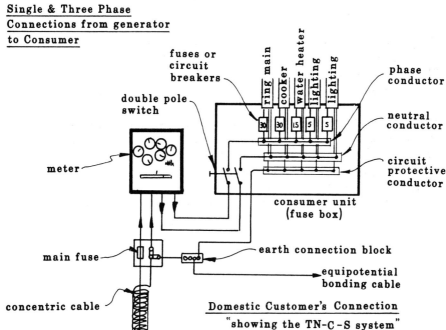

fuses or
circuit
breakers

ring main
cooker
water heater
lighting
lighting

phase
conductor

double pole
switch

neutral
conductor

30 30 l5 5 5

meter

circuit
protective
conductor

consumer unit
(fuse box)

main fuse

earth connection block

equipotential
bonding cable

concentric cable

Domestic Customer's Connection
"showing the TN-C-S system"

Electrical Supply

Electrical Safety

Relevant British Standard
BS 7671

It is a requirement that no person should carry out any electrical work unless competent to do so. This small introduction to electrical installation work is designed to give only a basic knowledge of electrical work.

Circuit protection

When a current in excess of that required passes down a conductor (cable) due to an overload or a fault developing, causing a short circuit, it is essential that the supply is automatically cut off. The Institution of Electrical Engineers (IEE) stipulates that the current is disconnected automatically within 0.4 seconds should a greater current flow through a conductor than that for which the circuit was designed. The time is increased to 5 seconds for circuits with fixed appliances, providing they are not located in a bathroom or shower-room. Several devices are used to detect such overloads, including the fuse and the miniature circuit-breaker (mcb).

The **fuse** is simply a short length of wire which will heat up and melt should the flow of electrons exceed its design ampere rating. The most common fuse is the cartridge type, although it is possible to still find rewirable fuses. The fuse rating is generally calculated by applying the formula: watts ÷ volts = amps; this formula does, however, ignore the resistance of the supply cable.

It is worth noting that the cartridge fuse fitted into a 13 amp plug feeding an appliance is often overrated. For example, a 100 watt light bulb only requires a fuse of $100 \div 240 = 0.4$ amps, to which a 1 amp fuse should be fitted; however, 1 amp is more than sufficient to kill at 240 volts. Usually, for the sake of convenience, a 3 amp fuse, which may be purchased at the local shop, is fitted. It may not blow immediately but the cable used is protected, this being the prime function of the fuse. The earth is to protect the user! Equally a 13 amp fuse should not be fitted because it is unnecessarily high.

The **mcb** is a device which usually includes a temperature-sensitive element to operate an electromagnet, causing it to trip a switch which closes, breaking the flow of current. They are more expensive than fuses but have the advantage of being re-setable – and they can be used as a switch to control the circuit they serve.

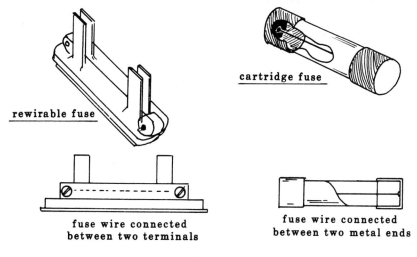

rewirable fuse

cartridge fuse

fuse wire connected
between two terminals

fuse wire connected
between two metal ends

Fuses

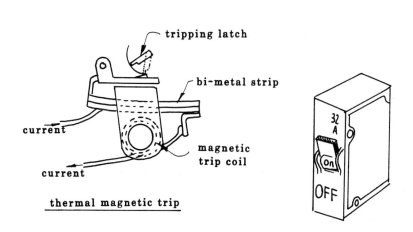

tripping latch

bi-metal strip

current

current

magnetic
trip coil

thermal magnetic trip

32
A

on

OFF

Miniature Circuit Breakers

Electrical Safety

Earth Continuity

Relevant British Standard
BS 7671

Equipotential bonding It is a requirement of the IEE Regulations that all exposed metalwork within a building be bonded together and connected to the main earthing block at the consumer unit. In the case of any metallic domestic services entering a building (e.g. water, gas or oil) the buried pipe itself may provide a path for any stray electrical currents; however, the current may cause corrosion to the pipework; also, should someone disconnect any part of the service pipework he/she might receive an electric shock. To prevent this the supplies are bonded together at their point of entry to the building and, in the case of the gas supply, within 600 mm of the gas meter.

The main equipotential bonding conductor should have a cross-sectional area of not less than 10 mm^2 and be run to all extraneous conductive parts (i.e. any exposed metal which is not part of the electrical installation), including accessible structural steel members, main ventilating ductwork, lightning conductors and central heating pipes, as well as the main services.

Supplementary bonding The main equipotential bonding connects only to one point of the metalwork. To ensure that earth continuity is maintained throughout the system additional cross-bonds are required. For example, the equipotential bonding wire connects to the cold supply pipe as it enters the building, but because the storage cistern, if applicable, may be made of plastic or because of the method of jointing the pipework (e.g. plastic fittings), the earth continuity may not be maintained throughout the hot and cold pipework. Equally at stainless steel sinks, etc., a good metallic bond may not have been achieved.

Supplementary bonding wires, usually with a cross-sectional area of 6 mm^2, are used to join these parts together, thus ensuring that whatever metalwork is touched a good continuity is maintained with the earth. If earth continuity tests prove that the metalwork is effectively bonded, supplementary bonding may not be necessary.

Earth clips When securing the bonding wire to pipework a special clip should be used (see figure). Note that the cable is hooked onto the clip and tightened, leaving a section of insulation material intact. Where a conductor is to loop to another service pipe the cable should not be cut. On making the connection, ensure that a good contact is made with the metalwork.

Temporary bonding wire At any time, should a section of metalwork be removed (e.g. a section of pipework), it is essential that the earth continuity is maintained before the disconnection occurs. This is achieved by bridging the section with a 10 mm^2 piece of cable. Thus, if there is a fault on the system and you touch live metal while a live current is passing down to earth, you are prevented from getting an electric shock because, should disconnection with the earth route via the metalwork occur, the current will flow through you.

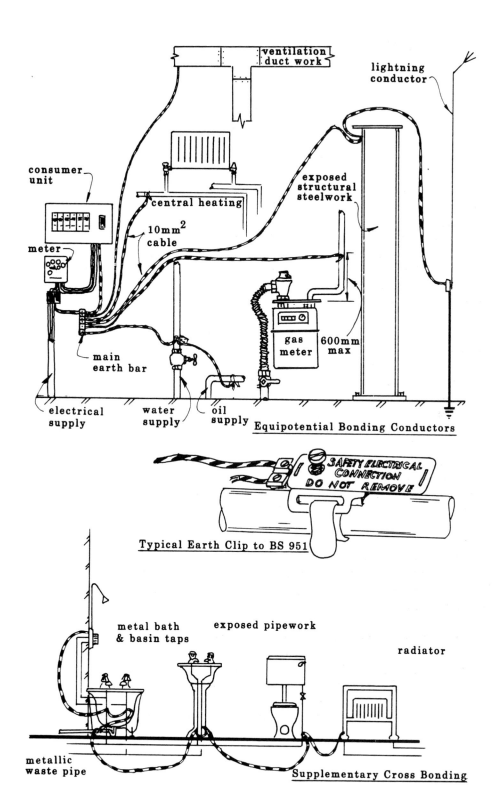

Equipotential Bonding Conductors

Typical Earth Clip to BS 951

Supplementary Cross Bonding

Earth Continuity

Domestic House Wiring 1

Relevant British Standard
BS 7671

In general three types of wiring circuits will be found in domestic installations: the supply to the 13 amp socket-outlets; the lighting circuit; and the various radial circuits to fixed appliances.

13 amp socket-outlets

13 amp socket-outlets are either fed from a continuous ring circuit or a radial circuit. The *ring circuit* is the most common system in which the phase, neutral and circuit protective conductor (cpc) are connected to their respective terminals in the consumer unit and from here the cables circulate from one socket to the next, passing round to all the sockets in one big loop, and returning to the consumer unit.

There is no limit to the number of outlets served by the ring main although the maximum floor area, in a domestic situation, served by a single 30 amp fuse or mcb must not exceed 100 m^2. The *radial circuit* differs in that the final socket on the system does not feed back to the consumer unit and as a result the maximum floor area to be served must not exceed 50 m^2 (i.e. half that of a ring circuit). The size of the conductor used is usually 2.5 mm^2 twin and earthed PVC insulated cable.

Spur outlets Where it is inconvenient to incorporate a socket-outlet within the loop of the circuit, such as an isolated location, or an addition, it is possible to run a single twin and earth wire to the socket. The spur is connected to the circuit, usually via a joint box or directly from the back of an existing socket. If the spur is fused an unlimited number may be connected to the circuit; however, where not fused the total number of spurs must not exceed the total number of socket-outlets connected directly to the circuit. A non-fused spur may supply either a single or double outlet. The cable size for a spur should be equal to that of the main circuit.

Lighting circuit

The lighting system is a radial circuit which feeds each ceiling rose or wall light in turn as it passes round the system. To prevent the light being continuously *on* the phase conductor does not pass directly to the bulb but via a switch, usually mounted at the entrance to the room at about 1.5 m above the floor level.

The circuit is protected with a 5 amp fuse or mcb and the usual size of conductor is 1.5 mm^2 twin and earth PVC-insulated cable. Generally a one-way switch is required to turn on and off the light; however, a two-way switch is used on stairways, etc., so that light can be operated from more than one location. The two-way switch requires a special switch control and is wired up as shown. Due to the size of the fuse and to prevent nuisance tripping when many lights are on, often the house is divided into two circuits (e.g. one up and one down).

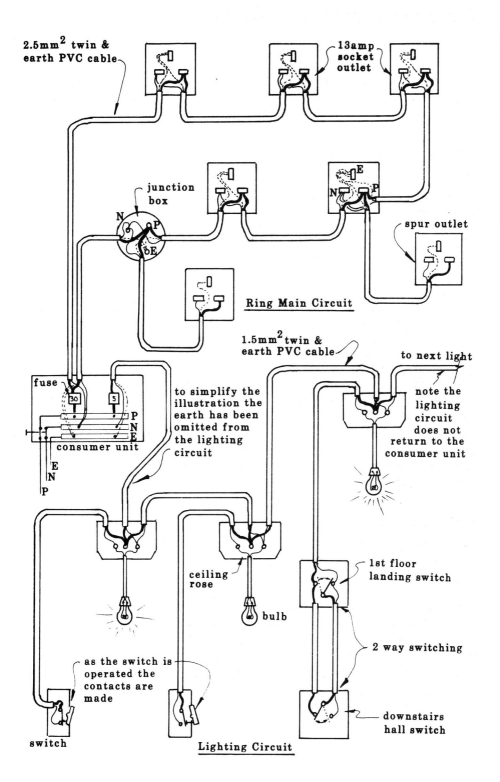

2.5mm^2 twin & earth PVC cable

13amp socket outlet

junction box

spur outlet

Ring Main Circuit

1.5mm^2 twin & earth PVC cable

to next light

note the lighting circuit does not return to the consumer unit

fuse

to simplify the illustration the earth has been omitted from the lighting circuit

consumer unit

ceiling rose

1st floor landing switch

2 way switching

as the switch is operated the contacts are made

bulb

switch

downstairs hall switch

Lighting Circuit

Domestic House Wiring 1

Domestic House Wiring 2

Relevant British Standard
BS 7671

Radial circuits to fixed appliances

Various radial circuits will be found in a home, including the supply to an immersion heater, a shower heater and a cooker control unit. The phase, neutral and cpc are run to terminate close to the appliance with a double pole switch (i.e. both phase and neutral are switched). From this point the supply is run to the appliance as necessary.

The power supply to an immersion heater This is a typical radial circuit in which 2.5 mm^2 twin and earth PVC-insulated cable is run from the consumer unit, the phase supplied via a 15 amp fuse or mcb, to terminate at a 21 amp double pole switch with neon indicator. From this point a 21 amp heat-proof flex is run to the immersion heater element, the phase being supplied via the thermostat as shown in the figure.

Fixed appliances such as a boiler must be wired up so that when the power is switched off and no longer feeds the appliance both phase and neutral conductors are disconnected. This is achieved usually using a double pole fused spur box, although it is possible to use an *unswitched* socket outlet; thus, to ensure that the power is off the plug must be removed.

Switches in bath and shower rooms No 13 amp sockets are permitted in these rooms and where switches are required to operate lights, pumps and heaters, etc. these must be via a pull cord located at high level.

Termination of a 13 amp three pin plug

The wires to be terminated into a plug need to be securely anchored in position. The earth must always be the conductor which would be the last to pull from its connection if the cable was to be pulled; thus the earth continuity is maintained for as long as possible. The wires into the plug are connected as indicated in the following chart. **Note**: although the plug is referred to as 13 amp, this does not mean that the fuse to be fitted should be of this size (see page 224, Electrical Safety).

Conductor	New colour code	Older appliances	Notes
Phase	Brown	Red	To connect via the fuse
Neutral	Blue	Black	
Earth	Green & Yellow	Green	To the largest pin

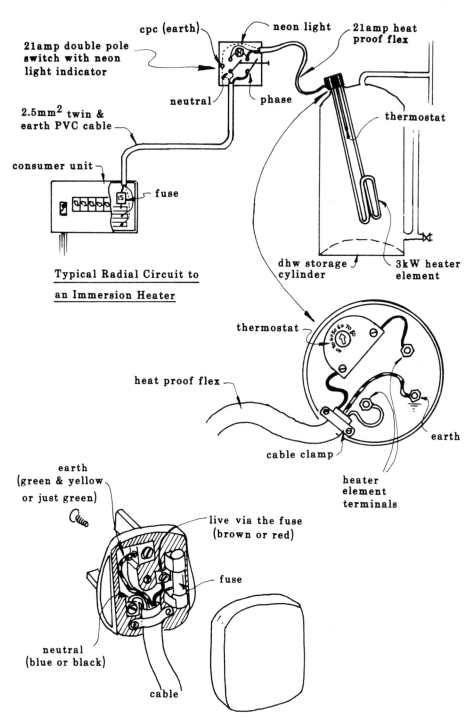

cpc (earth)

neon light

21amp heat proof flex

21amp double pole switch with neon light indicator

neutral

phase

thermostat

2.5mm^2 twin & earth PVC cable

consumer unit

fuse

Typical Radial Circuit to

an Immersion Heater

dhw storage cylinder

3kW heater element

thermostat

heat proof flex

earth

cable clamp

heater element terminals

earth (green & yellow or just green)

live via the fuse (brown or red)

fuse

neutral (blue or black)

cable

Termination of cable to a Plug

Domestic House Wiring 2

Installation Practices

Relevant British Standard
BS 7671

All electrical installation work must comply with the Electrical Supply Regulations 1988. Compliance with the 16th edition of the Institution of Electrical Engineers Regulations (IEE Regs) generally covers all aspects of the statutory regulations in force. The IEE Regulations are not legally enforceable, but in following their requirements you are deemed to be following the law. The electricity supply authority will not provide a supply of electricity until the conductors and apparatus are of sufficient size, and, as far as practically possible, constructed, installed and protected against danger and damage. The following identify just a few of the requirements to be met:

- All cable used is to be suitably insulated; usually for domestic circuits PVC twin and earth cable is used.
- The cable must be suitably supported and installed so that it is not liable to damage.
- Cables passing through holes in metalwork should be bushed using rubber grommets.
- Where the cable terminates it must be securely anchored; the cpc (earth) must be the last conductor to be disconnected should the supply cable suffer undue stress.
- Where the cpc is exposed it should be sleeved using green and yellow striped insulating material.
- A means of isolation to appliances must be provided in readily accessible locations.
- All circuits must be suitably protected with a fuse or circuit-breaker so as to prevent the risk of overloading the cable.
- Where the cable is concealed and installed in cement or plaster it should be protected against damage by covering it with a plastic or metal channel; alternatively, the cable may be run in conduit, suitably bushed at each end.
- If the cable is to be run under wooden floors or above ceilings the cable should be fixed to the side of the joists and where it is necessary to pass through a joist a hole must be drilled at least 50 mm from the surface to prevent damage caused from screws and nails securing any boards.
- Care should be taken to ensure that the cable is not allowed to come into contact with any metalwork such as gas or water pipes.

Termination of cables Where a cable is run into an appliance or accessary, such as a 13 amp socket, the cable must be securely anchored in position. It is, however, essential to ensure that the wire is not so taut as to prevent movement. The reason for this is that when the current flows through the conductor a certain amount of heat is generated, causing the conductor to expand and contract; this movement, if restricted, will cause the conductor to pull out from the point of termination.

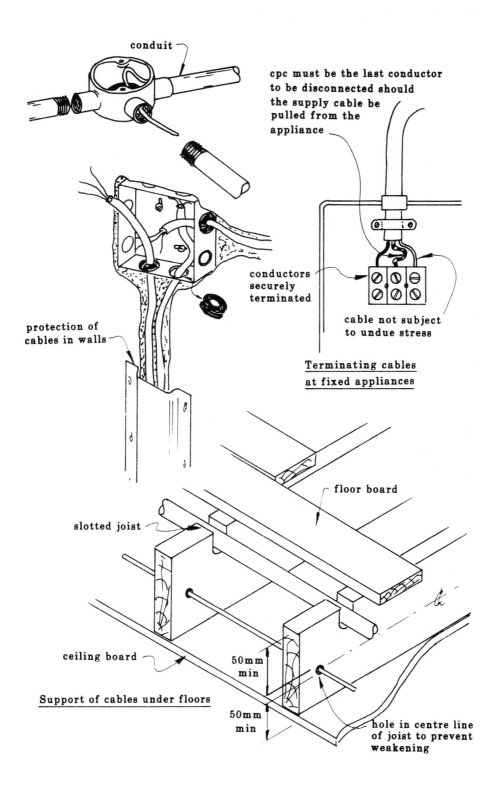

conduit

cpc must be the last conductor
to be disconnected should
the supply cable be
pulled from the
appliance

conductors
securely
terminated

cable not subject
to undue stress

protection of
cables in walls

<u>Terminating cables
at fixed appliances</u>

floor board

slotted joint

ceiling board

50mm
min

<u>Support of cables under floors</u>

50mm
min

hole in centre line
of joint to prevent
weakening

Installation Practices

Electrical Components 1

Thermostats

Thermostats are devices which detect a change in temperature and make or break the contacts of a switch. Thermostats generally comprise one of the following:

Bi-metallic strip: Usually found in room thermostats. It consists of two metal alloys, such as brass and Invar, which have different expansion rates, secured together. As the environment heats up, one of them (the brass) will want to expand; however, the other (the Invar) will not; this causes the bimetal strip to bend breaking the electrical contact.

Differential expansion: Typically found in an immersion heater. Again it consists of two metal alloys with different expansion rates which are secured at one end (see figure). As the brass tube heats up it expands, pulling with it the Invar rod, onto which is attached the moving contact. On cooling, the brass contracts to remake the connection. Note that the brass tube does not actually touch the water but sits into a pocket in the heater element.

Liquid expansion: Most commonly found in boilers in which the sensor sits into a pocket at the top of the heat exchanger. A volatile fluid, such as mercury or alcohol, is enclosed in a sensor (phial) and capillary tube which is connected to a bellows. As the fluid heats up it expands into the bellows chamber which fills and becomes enlarged; this in turn acts upon the push rod which causes the electrical contact to disconnect.

Ignition systems

In order to light the fuel of gas- and oil-burning appliances some form of ignition is required, often automatically. The ignition of a flame, usually a pilot flame in the case of gas supplies, can be achieved by either filament or spark ignition.

Filament ignitors: A small coil of thin resistance wire, usually of platinum. When a power source is supplied, at around 3 V, to the coil the wire heats up and glows red hot. The power may be supplied by a battery or a step-down mains transformer.

Spark ignitors: Several designs of spark ignitor will be found, including the piezoelectric device and the step-up transformer.

With the *piezoelectric device* two crystals, e.g. quartz, each about 6 mm in diameter and 12 mm long, are positioned into a metal holder and separated by a metal pressure pad. This pad is connected to the spark electrode and the holder suitably earthed. When pressure is applied to the crystals, usually by the operation of an impact spring device, an emf of about 6000 V is produced; this causes a spark to jump between the electrode tip and the metalwork of the appliance.

The *step-up transformer* is used to produce an emf of around 5000–15 000 V, which causes a series of electrical impulses to discharge a spark across the gap between two electrodes, or one electrode passing to earth. Systems of this nature are used to ignite non-permanent pilots and pressure jet oil burners.

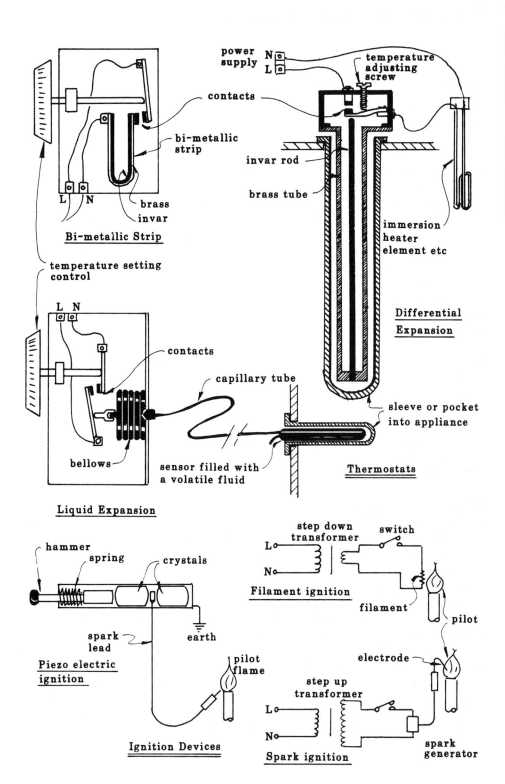

power supply N L

temperature adjusting screw

contacts

bi-metallic strip

invar rod

brass tube

brass
invar

L N

Bi-metallic Strip

immersion heater element etc

Differential Expansion

temperature setting control

L N

contacts

capillary tube

bellows

sensor filled with a volatile fluid

sleeve or pocket into appliance

Thermostats

Liquid Expansion

hammer
spring
crystals

spark lead
earth

Piezo electric ignition

pilot flame

Ignition Devices

step down transformer   switch

L
N

Filament ignition

filament
pilot

electrode

step up transformer

L
N

Spark ignition

spark generator

Electrical Components 1

Electrical Components 2

Relevant British Standard
BS 7671

Transformers Devices which step up or down the voltage from an a.c. supply. When a coil of wire is wound onto a soft iron core and an electric current passed through the wire a magnetic field is generated. If the voltage is a.c. the current will flow back and forth, and cause the lines of magnetic flux also to change direction. When a magnetic field is passed over a coil of wire the current is induced onto the wire conductor (see page 220 on electrical current referring to a.c.). Therefore, if a second wire is also wrapped onto the soft iron core the electric current will be induced from one coil (the primary coil) to the other (the secondary coil). The amount of turns on the secondary coil determines whether the voltage will be stepped up or down.

Example: where half the number of turns are on the secondary coil the voltage will be halved; conversely, where twice the number of coils are made the voltage will be doubled. Where a 110 V transformer is used on site a centre tapping has been made into the secondary coil and run to earth. Under normal operations 110 V will be supplied; however, in the event of a fault only 55 V will travel down the conductor to the appliance and its user and 55 V will run down to earth.

Solenoid valve (magnetic valve) A device used to hold a valve open, typically found in washing machines and gas valves. The solenoid consists of a coil of wire wrapped round a soft iron core. When a current of electricity is passed through the wire it creates a magnetic field which will in turn attract a soft iron armature, drawing it up into the coil. Attached to the armature is a valve; thus, when energised the valve will open. When the current flow is switched off the valve, assisted by a spring, closes.

Relay The relay is a device which uses the current from one circuit (e.g. low voltage) to switch on the current to another circuit (e.g. mains voltage). The relay consists of a solenoid which, when energised, draws in the armature to cause two contacts to be made on a switch. Relays will be found in many situations such as when switching the mains voltage to a pump from a low voltage thermostat circuit.

Rectifier (a.c./d.c. convertor) A device which converts a.c. to d.c. It is like a non-return valve allowing current flow only in one direction. When a single diode rectifier is fitted in a conductor the current will flow intermittently (i.e. stop and start) due to the fact that a.c. voltage flows back and forth between phase and neutral. However, when a bridge rectifier is used the current is allowed to flow continuously due to its design containing four single rectifiers. Rectifiers are found in many appliances such as transistor radios which work from a d.c. supply.

Capacitor (condenser) A device which stores a charge of electrical energy. The capacitor consists of two sheets of metallic foil separated by a thin sheet of insulation material, usually wrapped in a roll for convenience. When an a.c. voltage is fed to the capacitor the electrons, when flowing in one direction, enrich one metal foil; when the current is flowing in the opposite direction the second foil becomes enriched with electrons, whilst at the same time the first foil discharges its previous positive charge. The process continues as long as the current flows. This, in effect, creates an out-of-phase differential which is like having a two-phase supply.

A capacitor is typically found fitted to small motors, as used in c.h. pumps to give the initial impetus to get the motor going. It must be borne in mind that when the current stops flowing the capacitor still has a positive charge of electricity. Therefore, when removing a capacitor the terminal connections should be earthed to remove the charge.

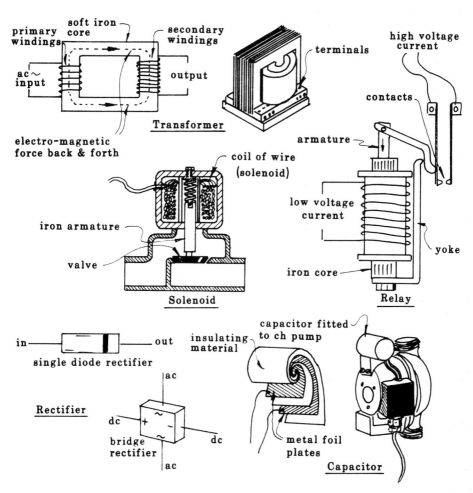

Electrical Components 2

Central Heating Wiring Systems

When a c.h. system has been completed it will need to be wired up in accordance with the manufacturer's instructions. It is not necessary to know the installation function of the various elements or how the circuit operates to make the required electrical connections. Manufacturers go to great lengths to design their wiring systems in such a way as to ensure that an installation can be achieved by anyone competent to do so. Basically all one has to do is follow a series of steps, as laid down by the manufacturer, which identify what cable is run where.

For example, in order to complete a c.h. system which is fully pumped to the dhw and c.h. via a mid-position diverter valve (using for argument's sake the Honeywell controls listed below) the wiring would be completed by running all the cables to the numbered location in a strip connection block, as identified below:

Cable from the Component	to Connection Block No	Cable from the Component	to Connection Block No
Mains input		**Cylinder thermostat Honeywell L641A**	
phase	1	*Terminal*	
neutral	2	1	8
earth	3	2	7
		C	6
Boiler and pump		**Room thermostat Honeywell T6360B**	
phase	8	*Terminal*	
neutral	2	1	4
earth	3	2	2
		3	5
Mid-position valve Honeywell V4073A		**Programmer Honeywell ST 6300**	
Cable		*Terminal*	
White	5	1	7
Blue	2	3	6
Grey	7	4	4
Orange	8	N	2
Green & Yellow	3	L	1

Notes: The boiler and pump would differ from one make to another. I have simply chosen a basic boiler. The cables should be run, complying with IEE standards, to the terminal strip and completed as shown in the wiring figure opposite. It is not possible to identify all the various wiring plans available; however, a manufacturer of these controls, such as Honeywell or Landis & Gyr would be happy to provide a simple-to-follow guide.

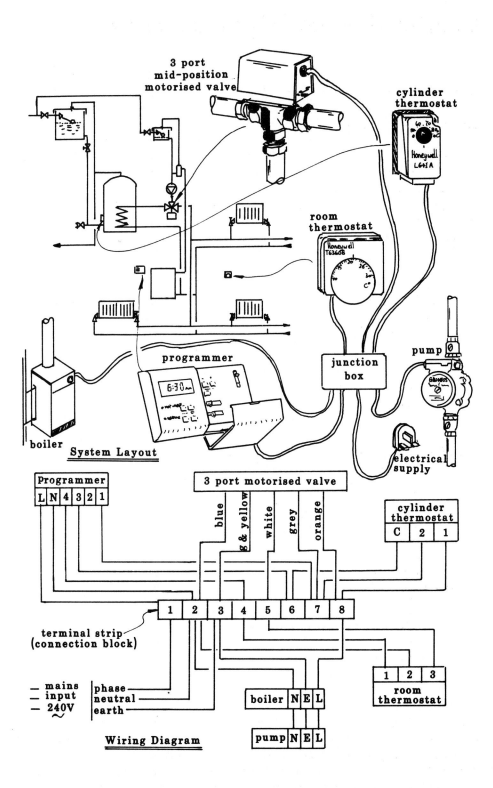

Central Heating Wiring Systems

Inspection and Testing of Electrical Work

Relevant British Standard
BS 7671

Before carrying out any work on electrical circuits it is essential that the electrical supply is isolated. The main mcb or isolating switch should be capable of being locked open, breaking the circuit or any fuses should be removed.

A visual inspection should be made of the system to ensure that all connections are properly secured into the terminal connections and that all components are securely fixed. A check should be made of the fuse connected to ensure it is not overrated (see page 224). The plumber is generally only interested in testing earth continuity and polarity of wires at an appliance.

Testing for earth continuity

This is a test carried out to ensure that any exposed metalwork (e.g. appliances such as boilers, exposed pipework, sink tops, etc.), is effectively connected to the main earth terminal block. To carry out this test either an ohmmeter or multi-meter set to a low resistance setting (Ω) is used. Note it is possible to carry out this test using a circuit tester with a buzzer.

The procedure to follow is to position one probe on to the exposed metalwork and the other probe on to a known earth (e.g. the metal screw exposed on the surface of a 13 amp 3-pin socket-outlet). The reading on the dial should be less than 1, which indicates a very low resistance to current flow and ensures a good earth. It may be necessary to use a cable to extend the length of the test leads; where this is the case the resistance of the test lead should be taken, then subtracted from the final reading.

It should be noted that the method described here confirms a good earth continuity but does not say how good the earth connection is, i.e. it may only be one thin strand of copper. This connection should be physically/visually inspected.

Testing for correct polarity

This is a test done to ensure the phase conductor is connected to its correct location at the appliance and not crossed somewhere, which may result in the appliance casing becoming live.

Polarity test with the power off Achieved by testing the continuity of the conductor from its source to the appliance, using a low resistance (Ω) setting on the meter. The test is similar to the earth continuity test. One probe of the multi-meter is positioned on to the phase conductor of the fused spur box and one on to the connection of the appliance; the reading should be *less* than one. Keeping the probe in position at the spur the second probe is now repositioned on to the neutral and then the earth terminals, respectively; in both cases the readings should indicate one. Testing for polarity with the power off is certainly safe; however, it does not confirm that the spur box, or source, is itself wired up correctly.

Polarity test with the power on Note that it is a statutory requirement that no one should work on live electrical equipment unless the work cannot be carried out in any other way and in all cases unless a person is competent to do so; when carrying out this test it is imperative that this requirement be observed.

Firstly with the power *off* remove the terminal cover from the appliance and disconnect or remove any integrated circuits. Set the multi-meter onto the a.c. selector to a reading position in excess to 240 V. Having turned the power on, position one probe on to the phase terminal and one on to the neutral, the reading on the dial will indicate the voltage being supplied, say 230 V. Leaving one probe on the phase, position the other on the earth terminal the voltage should again read that being supplied: 230 V. Finally position one on the neutral and one on the earth terminal to give a final reading, which should be zero (1–2 V may be showing, being a little residual current). Turn off the supply and replace the terminal cover, etc.

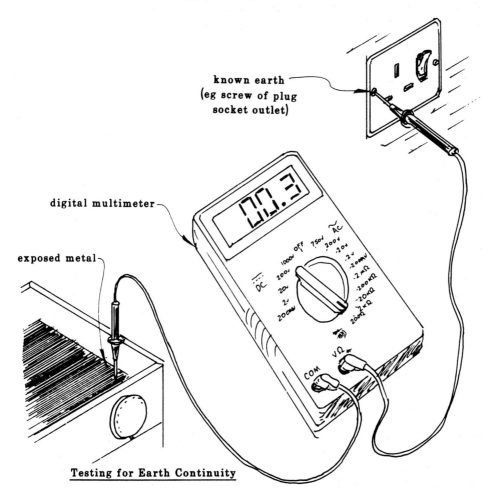

known earth
(eg screw of plug
socket outlet)

digital multimeter

exposed metal

Testing for Earth Continuity

Inspection and Testing of Electrical Work

Part 7
Sanitation

Sanitary Accommodation

Relevant British Standard
BS 6465

Sanitary accommodation is the term used for a room containing a water closet (WC) or urinal; it may or may not contain other sanitary fittings. The room is used solely for ablutionary (washing) and/or excretory purposes. The sanitary accommodation should not be entered directly from either:

☐ a room used for the preparation of food,
☐ an office or working area,
☐ a habitable room, except a bedroom, where additional sanitary accommodation is provided elsewhere in the dwelling.

Part G of the Building Regulations identifies the requirements for sanitary accommodation; in general every new dwelling should be provided with at least one WC; one bath or shower; one wash basin and one sink. The appliances should be suitably located, usually within a bathroom in which the bath or shower and wash-basin are fitted, with the WC in the same or adjoining room, thus allowing the WC compartment direct access to the wash basin. Where the dwelling is for more than five people it is recommended that a second WC compartment be installed. All sanitary appliances must be installed to a suitable drainage discharge system and connected to the drain via a trap. In buildings other than simple domestic dwellings the requirements regarding the number of rooms for sanitary accommodation are clearly identified in BS 6465 which should be sought for further study.

Ventilation requirements These are identified in the Building Regulations under Part F in which all sanitary accommodation should have **either**:

(1) A window, skylight or some other suitable means of ventilation opening directly to the external air, with a total area capable of being opened of at least $\frac{1}{20}$th of the floor area, or
(2) A means of mechanical ventilation, capable of extracting air to give not less than three air changes per hour. This is usually achieved by the operation of the light switch with a 15 minute overrun to the extractor motor.

Spatial requirements The internal layout and arrangement of the sanitary fitments and facilities should be determined by:

☐ The dimensional characteristics of the human body, allowing for an activity space, in which there is room to move;
☐ The number of persons using the facility;
☐ The size of each component, the associated services and their location in relation to the component. Also, provision should be made to gain access, especially to drainage, for servicing and maintenance purposes.

7 Sanitation

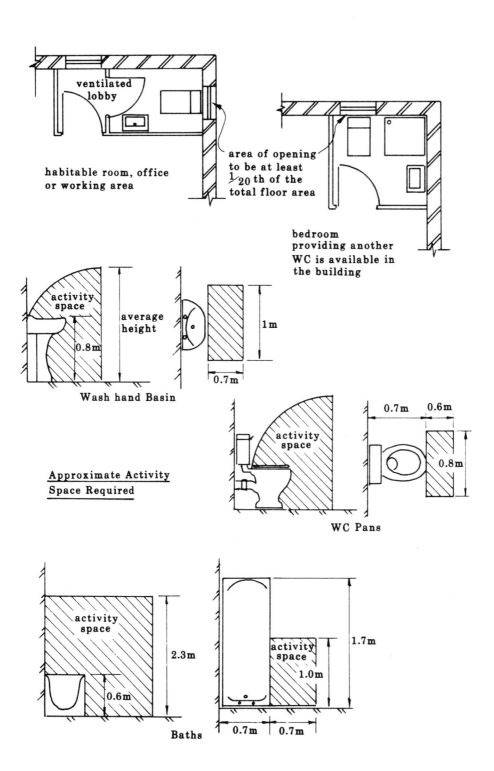

ventilated lobby

habitable room, office or working area

area of opening to be at least $\frac{1}{20}$ th of the total floor area

bedroom providing another WC is available in the building

activity space

average height

0.8 m

1 m

0.7 m

Wash hand Basin

Approximate Activity Space Required

activity space

0.7 m   0.6 m

0.8 m

WC Pans

activity space

2.3 m

0.6 m

Baths

activity space

1.7 m

1.0 m

0.7 m   0.7 m

Sanitary Accommodation

Sanitary Appliances 1

Sanitary appliances can be broadly divided into two groups: those which are used for washing purposes, i.e. waste or ablutionary appliances; and those used for the removal of human excreta, i.e. soil appliances.

Waste appliances

Sinks Several designs of sink will be found. Firstly there are those fixed at a height not exceeding 900 mm or level with adjoining storage cupboards and used for washing items. These include stainless steel and enamelled pressed steel crockery sinks and glazed fireclay butler sinks.

The most common design of butler sink is the 'Belfast' design, having an integral overflow; this kind of sink is mostly found in utility rooms. A second type of sink is the bucket or cleaner's sink which has a protective strip fitted to its front edge and is provided with a hinged metal grating, on which to rest a bucket. These sinks are fitted at low level to facilitate the emptying of buckets.

Wash basins These are bowl-shaped fittings used for the purpose of washing the upper parts of the body; they are usually made from vitreous china. The fixing height of a basin is usually determined by a pedestal or vanity unit although where brackets are used it should be fixed at a height of 760–800 mm.

Bidets These are appliances designed for the purpose of washing the lower parts of the body. Bidets are made from the same materials as wash basins and are usually floor standing. Two designs of bidet will be found: those with an over-rim water supply and those with a submersible ascending spray (see page 94, Connections to Hot and Cold Pipework).

Baths Most baths installed today are made from acrylic materials, for which adequate support is essential. The bath is intended to allow complete bodily immersion in the reclining position, although it is possible to have a bath with a stepped bottom (sitz bath) where floor space is restricted or for the disabled person. A bath is generally fitted as low as possible, its height usually being determined by the height of the accompanying side panel.

Showers These are sometimes incorporated as part of a bath; however, it is possible to purchase shower trays to give maximum usage of floor space, the tray forming part of a shower cubical.

In general all appliances should be securely fixed using the brackets supplied. The support of service and waste connections should not be relied upon. The appliance should be levelled in along its top edge; the fall is built in with its design. The size of the waste pipe connections is identified on page 252.

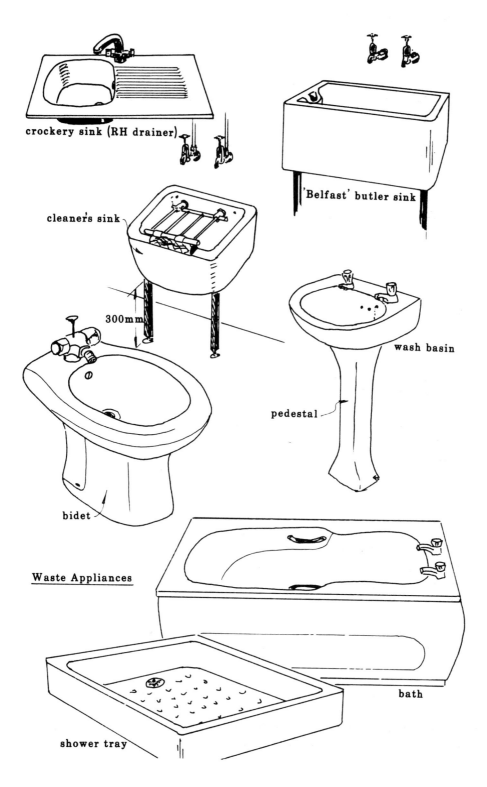

crockery sink (RH drainer)

'Belfast' butler sink

cleaner's sink

300mm

wash basin

pedestal

bidet

Waste Appliances

bath

shower tray

Sanitary Appliances 1

Sanitary Appliances 2

Relevant British Standard
BS 6465

Soil appliances

WC pan An appliance used in conjunction with a flushing cistern to remove the contents of solid and liquid excreta from the building by the flush of water. The pan may be floor standing or mounted on the wall, leaving the floor area clear. Two common designs of WC will be encountered: the *washdown* and the *siphonic* pan.

The washdown pan relies solely on the momentum of the flushing water entering the pan from above to remove the contents of the bowl whereas the contents of the siphonic WC pan are removed by siphonic action in addition to the momentum of the falling water.

Two designs of siphonic pan will be found: the single trap and the double trap designs; in each case the principle is to create a negative pressure below the trap seal. This is achieved by restricting the outflow in the case of the single trap pan, or drawing the air from the void between the two traps in the case of the double trap pan. With the double trap pan the suction of air is achieved when the water from the flush passes over a pressure-reducing fitting.

Most WC pans have an outlet size of 100 mm although it is possible to find pans with a 75 mm outlet.

The slop hopper or slop sink An appliance, similar to a WC pan, with a flushing rim, supplied with water from a flushing cistern; it is used to discharge the contents of bed pans and urinal bottles. The difference between slop hoppers and slop sinks is simply that the hopper sits on the floor whereas the sink is installed at a height of about 1 m. In addition to the flushing cistern these appliances are installed with hot and cold supplies fed to bib cocks located above the appliance. The outlet size is 100 mm diameter.

Urinals Appliances found only in male toilets for the removal of urine. Four designs of urinal will be found: the bowl, trough, slab and stall. Bowl and trough urinals are installed at a height of 610 mm to the front lip, 510 mm for junior boys. The stall and slab are floor-standing appliances, a channel being positioned at ground level, not exceeding 2.4 m to the discharge outlet.

Urinals are flushed periodically with a discharge of water from an automatic flushing cistern to allow the whole surface of the appliance to be cleaned. Bowl and stall urinals are fitted with a spreader at each location (see figure) whereas for trough and slab urinals a sparge pipe may be used, consisting of a perforated pipe which is the same length as the appliance. Thus a discharge of water along the whole length of the slab/trough back is facilitated. Alternatively spreaders may be fixed at 600 mm centres. To provide privacy to the user, divisions may be included with the installation allowing 600 mm per unit.

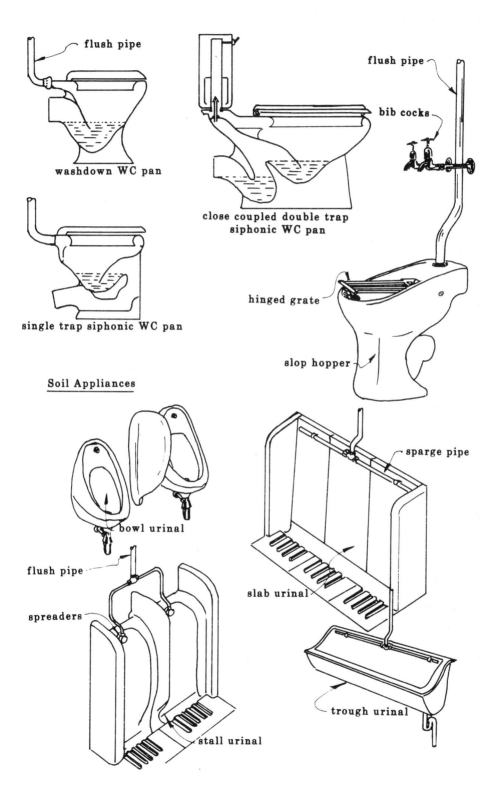

flush pipe

washdown WC pan

single trap siphonic WC pan

close coupled double trap
siphonic WC pan

flush pipe

bib cocks

hinged grate

slop hopper

Soil Appliances

bowl urinal

flush pipe

spreaders

sparge pipe

slab urinal

stall urinal

trough urinal

Sanitary Appliances 2

Flushing Cisterns

Relevant British Standards
BS 1125 and BS 1876

Plunger type flushing cisterns

A cistern which supplies a flush of water to a WC pan for the purpose of removing the contents of the bowl. Water is supplied to the cistern high in the appliance, the water level being controlled by a float-operated valve (ballvalve) to give a contents of 7½ litres. Note that prior to 1993 a 9-litre flush was employed; this is no longer permitted except for replacement purposes. Also no longer permitted are cisterns giving a dual flush; these allowed for a 4½ or 9 litre flush. When a flush of water is required the flushing arm is operated by pulling a chain on high-level cisterns or turning the handle on low-level cisterns, causing the plunger in the cistern to lift. This forces a flow of water to be discharged over the top of the U-shaped siphon bend; as the water falls down the flush pipe it carries with it the air from within the pipe, lowering its pressure.

The atmospheric pressure acting on the water in the cistern pushes down and forces all the water from the cistern by siphonic action, up and over the bend and down the flush pipe. When the water level drops to reach the base of the plunger, air can get in to equalise the pressure inside the tube and the siphonic action ceases. The cistern may be bolted directly to the WC pan; it may be part of a close-coupled suite or fitted with a flush pipe in which the cistern is fitted at either high or low level.

The diameters of the flush pipes should be 40 mm and 32 mm respectively for low- and high-level cisterns. Where a flush pipe is used it connects to the pan by the use of a plastic connector/cone. An overflow will need to be run from the cistern to terminate in a visible location, usually externally to the building.

Automatic flushing cistern

A flushing cistern designed to discharge its contents of water at regular intervals into a urinal. The rate at which the water will flush depends upon the rate at which the water is fed into the cistern and for a single installation this should not exceed 10 litres per hour.

Where several urinals are installed the filling rate should not exceed 7.5 litres/h per bowl, stall or 700 mm-width slab. To prevent the wastage of water from these cisterns, during times when the cistern is not used such as at weekends, an automatic flow cut-off device should be fitted.

The automatic cistern illustrated operates as follows:

(1) As the water rises upon filling, the air is trapped inside the dome; thus it becomes compressed and eventually when enough pressure is created the water is forced out of the U tube and the air pressure inside the dome is reduced.
(2) The reduced air pressure immediately allows siphonic action to start, thus flushing the appliance.
(3) When the flush is finished, the water in the upper well is siphoned out through the siphon tube and refills the lower well and U tube.

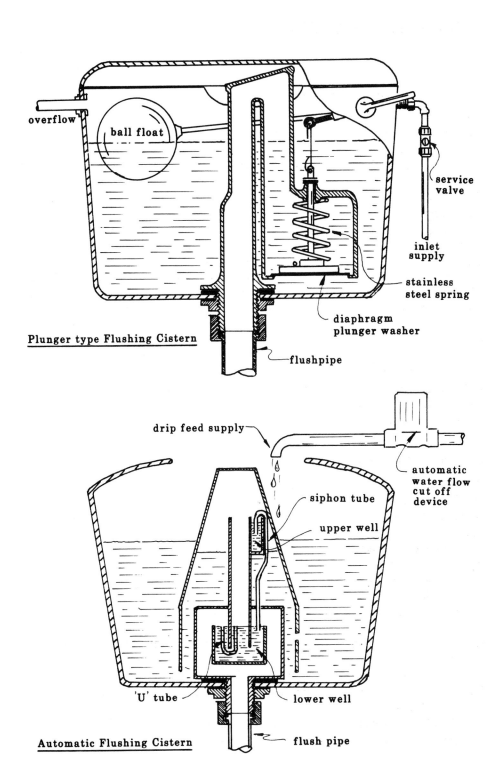

overflow

ball float

service valve

inlet supply

stainless steel spring

diaphragm plunger washer

Plunger type Flushing Cistern

flushpipe

drip feed supply

automatic water flow cut off device

siphon tube

upper well

'U' tube

lower well

Automatic Flushing Cistern

flush pipe

Flushing Cisterns

Waste Pipe Connections

Relevant British Standard
BS 5572

When an appliance has been installed it will need to be connected to the above-ground sanitary pipework. This is achieved by installing a waste fitting into the appliance and connecting it to the discharge pipe via a trap. The different minimum sizes of waste fitting and trap are identified in the table.

Minimum appliance waste sizes

Type of appliance	Waste fitting size (in)	Discharge pipe and trap size (mm)
Sinks, showers, baths, washing machines and trough urinals	1½	40
Wash basins, bidets, drinking fountains and bowl urinals	1¼	32
Stall and slab urinals	2½	65

Appliances such as sinks and baths use a combination overflow and waste fitting, which does away with the need for a separate overflow pipe discharging out of the building; note that the overflow connects into the waste above the trap seal.

Appliances such as the Belfast butler sink, wash basins and bidets have an integral (built in) overflow and therefore the waste fitting to be used needs to have a slot to allow its connection, whereas the urinal bowl has no overflow at all; thus an unslotted waste fitting must be chosen. Most waste fittings connected to the appliance are best fitted using a rubber-type material to give a good seal; however, materials such as Plumbers' Mait (a non-setting putty) can be used, providing the appliance is completely dry when it is applied.

Traps

A trap is a fitting or integral part of an appliance designed in such a way as to retain a body of water, thus preventing the passage of foul air. There are many different designs, but general trap designs include those with a vertical outlet (S traps) or a near-horizontal outlet (P traps), or the trap could be fitted in a pipe run in which case it would be called a running trap.

The depth of the trap seal would depend upon the circumstances and usage of the pipe, but in general pipes of less than 50 mm internal bore would have a trap with a seal of not less than 75 mm, although if the appliance is discharging into an open gully this depth could be reduced to 38 mm. For pipes with a larger internal bore than 50 mm a trap with a seal of 50 mm is required, the reason for this being that trap seal loss is much less likely to occur in a pipe so large.

Traps up to 40 mm outlet size may be either of a tubular or bottle trap design (see figure). Tubular traps tend to be less prone to blockage. (See page 260 for an example of a special trap which is designed to maintain its water seal under adverse design conditions.)

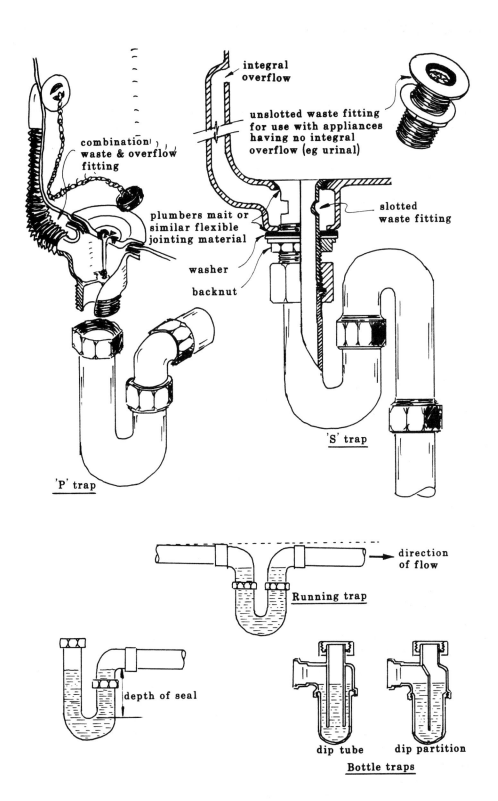

combination
waste & overflow
fitting

integral
overflow

unslotted waste fitting
for use with appliances
having no integral
overflow (eg urinal)

plumbers mait or
similar flexible
jointing material

slotted
waste fitting

washer

backnut

'P' trap

'S' trap

direction
of flow

Running trap

depth of seal

dip tube   dip partition

Bottle traps

Waste Pipe Connections

Sanitary Pipework

Relevant British Standard
BS 5572

A system of above-ground drainage designed to remove all the foul and waste water to the below-ground drainage system. Originally the foul water from soil appliances was kept separate from the water from waste appliances and two separate discharge stacks were required, the water only joined at ground level in the below-ground drainage system. This type of system was known as the two-pipe system.

All sanitary pipework today is based on the one-pipe system in which one discharge stack is used to convey both foul and waste waters. There are three basic systems in use, these being the *ventilated system*, the *ventilated stack system* and the *single stack system*.

With the **ventilated system** a ventilating pipe is extended to connect to each of the individual branch pipes throughout the system; it is designed to safeguard against trap seal loss (see page 260). This system is generally adopted in situations where it is not possible to have close groupings of sanitary appliances, and long branch discharge pipes can be expected.

In the **ventilated stack system** only the main discharge stack is ventilated, to over-come pressure fluctuations. With this system the branch discharge pipes connect directly into the main stack without the need for a branch ventilating pipe; this system, there-fore, is only suitable for buildings in which the sanitary appliances are closely grouped to the main stack.

Finally there is the **single stack system**, which is used in similar situations to the ventilated stack system, the difference being that the stack ventilating pipe can be omitted if the discharge stack is large enough to limit pressure fluctuations.

Whatever system is chosen all work must comply with Part H of the Building Regulations.

Branch discharge pipe This pipe should connect to the main discharge stack in such a way as not to cause any 'crossflow' into other pipes (see figure). The sizes of branch discharge pipes should be at least the same diameter as the appliance trap. Oversizing the pipe to avoid self-siphonage could prove uneconomical and lead to an increased rate of solid deposit accumulation. Bends should be avoided and where this is not possible long radius bends should be used. The gradient of branch discharge pipes should be between $1°$ and $1\frac{1}{4}°$ C (18–22 mm drop per metre run).

Branch ventilating pipe No branch ventilating pipe should connect to the discharge stack below the spillover level of the highest fitting served. The minimum size for a branch ventilating pipe serving a single appliance should be 25 mm, but, where the branch run is longer than 15 m or contains more than five bends or serves more than one appliance the minimum pipe size should be 32 mm.

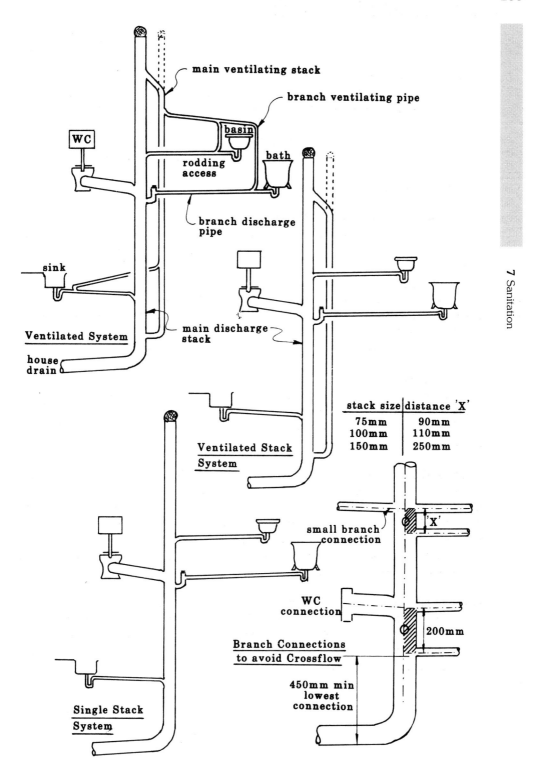

main ventilating stack

branch ventilating pipe

WC

basin

bath

rodding
access

branch discharge
pipe

sink

Ventilated System

main discharge
stack

house
drain

Ventilated Stack
System

stack size	distance 'X'
75mm	90mm
100mm	110mm
150mm	250mm

small branch
connection

'X'

WC
connection

200mm

Branch Connections
to avoid Crossflow

450mm min
lowest
connection

Single Stack
System

Sanitary Pipework

Single Stack System

Relevant British Standard
BS 5572

Most buildings are designed to meet the criteria of this system, therefore reducing the cost of installation. The system is designed so that no separate ventilating pipes are required to prevent trap seal loss. This can only be achieved by observing the following guidelines:

(1) All sanitary appliances must be closely grouped to the discharge stack, within the limits shown.
(2) All appliances, as far as possible, should be fitted with a P trap with a discharge pipe diameter equal to that of the trap and bends in branch pipes avoided – the gradient being kept to a minimum.
(3) The vertical discharge stack must be as straight as possible, with a long radius bend fitted at its base.
(4) The lowest connection to the discharge stack must be a minimum of 450 mm above the invert of the drain. If the building is over three storeys, this distance should be increased to 750 mm and for buildings over five storeys all ground-floor appliances should not connect into the stack. In buildings over 20 storeys all first floor, as well as ground floor, appliances should not being connected.
(5) Where a range of appliances is installed, it should comply with the following table:

Unvented discharge pipes serving more than one appliance

Appliance	Max no. fitted	Min pipe size (mm)	Gradient (mm/m)
WC	8	100	9–90
Wash basin	4	50	18–90
Bowl urinal	5	50	18–90
Stall urinal	6	65	18–90

(6) The main discharge stack must be large enough to limit pressure fluctuations without the need for a ventilating stack; as a general guide, 100 mm diameter is required for buildings with up to five storeys and 150 mm diameter for those with up to 20 storeys, with two groups of appliances on each floor. (**Note**: a *group* of appliances consists of one WC, bath, basin and sink.)
(7) Branch connections must join the main vertical discharge stack at an angle of 45° *or* at a radius of 25 mm for pipes up to 75 mm in diameter and a radius of 50 mm for pipes over 75 mm in diameter.

Where the discharge pipe exceeds the criteria listed it should be vented by a branch ventilating pipe, located at the highest point, extended to the atmosphere or connected to a ventilating stack, in which case the system is generally referred to as a 'modified single stack system'. Sometimes air admittance valves, or resealing traps, are used which allow air into the system in order to prevent trap seal loss.

Branch Connections		
pipe size	maximum length	gradient
32mm	1.7m	between 18-80mm/m
40mm	3.0m	” 18-90mm/m
50mm	4.0m	” 18-90mm/m
100mm	6.0m	minimum 18mm/m

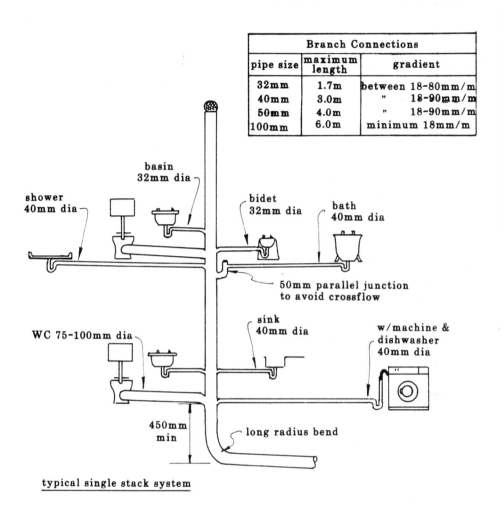

basin
32mm dia

shower
40mm dia

bidet
32mm dia

bath
40mm dia

50mm parallel junction
to avoid crossflow

sink
40mm dia

w/machine &
dishwasher
40mm dia

WC 75-100mm dia

450mm
min

long radius bend

typical single stack system

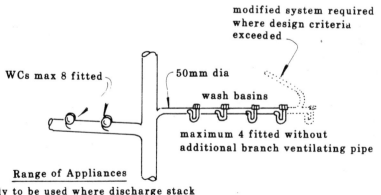

modified system required
where design criteria
exceeded

WCs max 8 fitted

50mm dia

wash basins

maximum 4 fitted without
additional branch ventilating pipe

Range of Appliances
(only to be used where discharge stack
is large enough to limit pressure fluctuations)

Single Stack System

Ventilation of Sanitary Pipework

Relevant British Standard
BS 5572

It is essential that air is allowed freely to enter discharge and drainage pipes and thus help maintain an equilibrium of pressure within the pipe and the outside atmosphere; for example, should the pressure be greater outside the pipe than inside, the trap seals of sanitary appliances and gullies would be lost. By allowing a current of fresh air to flow through the whole system, any foul matter adhering to the insides of pipes would soon dry and be washed away; ventilating the pipes also prevents any build-up of foul (and possibly dangerous) gases. Air enters the drain at low level, via holes in inspection covers, etc., and rises up through the stack by convection currents.

The termination of a ventilating pipe into the atmosphere should be at a position that does not cause a nuisance or health hazard. It is recommended that if a ventilating pipe is within 3 m of a window opening it should be carried up above the window to a minimum height of 0.9 m. Ventilating pipes should be fitted with a domical cage or grating which does not unduly restrict the free air flow and prevents the nesting of birds, etc. The size of the ventilating pipe may be reduced in size in houses up to two storeys, but should be at least 75 mm in diameter.

Discharge stacks may terminate within the building when fitted with an air admittance valve. These valves are designed to allow fresh air to enter the ventilating pipe but prevent odours and gases escaping. Air admittance valves should not be used on discharge stacks connecting to drains which are subject to surcharging (filling with water) or a drain which has an intercepting trap fitted, as it may result in the trap seal loss of the appliances.

The number of air admittance valves fitted to a drainage system should be limited to prevent excessive back pressures and where five or more domestic dwellings are located in the same drainage run a conventional open ventilating pipe will be required at the head of the system. Should the number of dwellings exceed ten, conventional venting will be required at mid-point and at the head of the system. Note that the valve must be located above the floor level of the discharge stack, and where they are fitted in roof voids, etc., they should be insulated, because any condensation forming within may freeze and prevent its use.

Stub stacks

A discharge stack which is capped off with a rodding eye at its top end. Stub stacks are only permitted to be installed where they connect to a ventilated discharge stack or drain within 6 m from the base of the stack. However, this distance is increased to 12 m maximum where a group of appliances is fitted. If a stub stack is used no branch waste connections may be made into the stack higher than 2 m above the invert level of the drain, and, in the case of a WC pan connection, this distance is reduced to 1.5 m maximum.

259

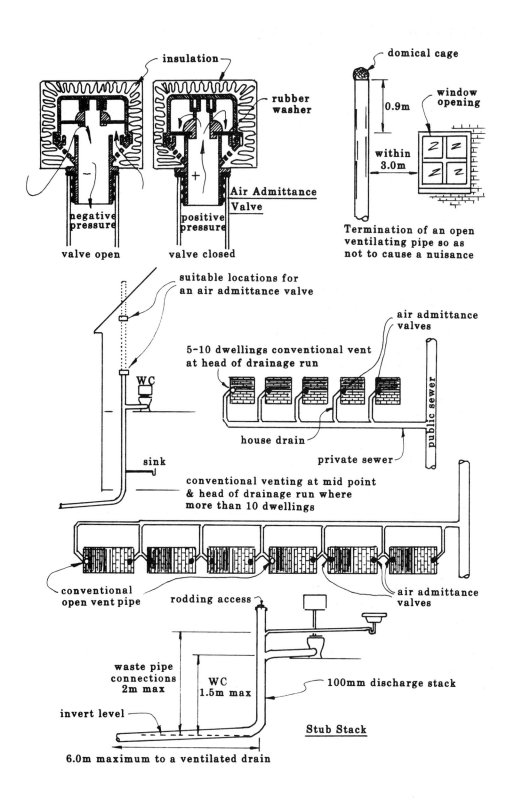

insulation

rubber washer

Air Admittance Valve

negative pressure

positive pressure

valve open

valve closed

domical cage

0.9m

within 3.0m

window opening

Termination of an open ventilating pipe so as not to cause a nuisance

suitable locations for an air admittance valve

WC

5-10 dwellings conventional vent at head of drainage run

air admittance valves

house drain

private sewer

public sewer

sink

conventional venting at mid point & head of drainage run where more than 10 dwellings

conventional open vent pipe

rodding access

air admittance valves

waste pipe connections 2m max

WC 1.5m max

100mm discharge stack

invert level

Stub Stack

6.0m maximum to a ventilated drain

7 Sanitation

Ventilation of Sanitary Pipework

Trap Seal Loss

Relevant British Standard
BS 5572

If the trap seal is lost objectionable smells will enter the building; therefore the water seal in the trap must be maintained under all circumstances. Trap seal loss can result from various unforeseen circumstances, such as leakages or evaporation. In designing any sanitary discharge system special care will need to be taken to prevent pressure fluctuations occurring within the pipework itself. Typical design faults include:

☐ **Waving out.** Caused by the effects of the wind passing over the top of the venti-lation pipe, bringing about pressure fluctuations; thus wave movements in the trap gradually wash over the outlet.

☐ **Compression.** This generally only occurs in high-rise buildings where the discharge of water down the main discharge stack compresses the air at the base of the stack, thus pushing the water out of the trap back into the appliance. This problem can usually be overcome by ensuring that a long radius bend is installed at the base of the stack and that no connections are made within 450 mm of the invert level of the drain; alternatively, a relief vent should be carried down to connect to the lowest part of the discharge stack.

☐ **Induced siphonage.** Caused by the discharge of water from another sanitary appliance connected to the same discharge pipe. As the water falls down the pipe and passes the branch pipe connected, it draws air from within, thus creating a partial vacuum; and subsequently siphonage of the trap takes place. To overcome this problem trap ventilating pipes could be designed into the system; these would permit air into the discharge pipe, preventing the development of a partial vacuum.

☐ **Self-siphonage.** Mostly caused in such appliances as wash basins; being funnel shaped they tend to discharge their contents of water quickly. As the water discharges it sets up a plug of water, which as it passes down the pipe creates a partial vacuum, thus causing siphonic action to take place. To overcome this problem of self-siphonage a larger waste pipe is sometimes used, but in most cases a resealing trap cures the problem.

Resealing traps

A trap designed to maintain its water seal should a partial vacuum be created in the waste discharge pipe. There are various designs of a resealing trap and the most common one used today incorporates an anti-vacuum valve; should the pressure drop inside the discharge pipe this valve opens under atmospheric pressure, giving a state of equilibrium inside the pipe. Unfortunately these traps often tend to leak through this valve. The other types of resealing trap work on the principle of retaining the water in a reserve chamber should the conditions be right for siphonic action to take place.

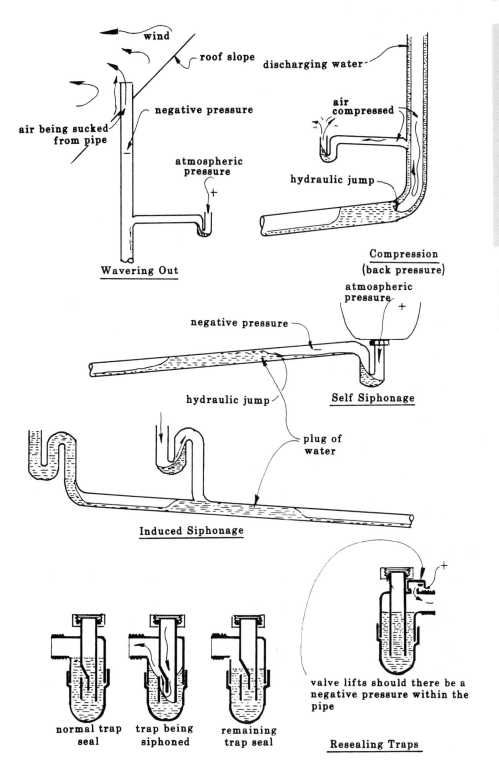

wind

roof slope

discharging water

negative pressure

air compressed

air being sucked from pipe

atmospheric pressure

hydraulic jump

+

Wavering Out

Compression
(back pressure)

negative pressure

atmospheric pressure

+

hydraulic jump

Self Siphonage

plug of water

hydraulic jump

Induced Siphonage

valve lifts should there be a negative pressure within the pipe

+

normal trap seal

trap being siphoned

remaining trap seal

Resealing Traps

Trap Seal Loss

Mechanical Disposal Units

Relevant British Standard
BS 3456

Food waste disposal unit

A mechanical device operated and fixed beneath sinks to macerate kitchen refuse into small fragments so that they can be flushed into the drainage system without causing blockages. A waste disposal unit cannot be fitted into a sink with the standard waste hole size as it would not be large enough to house the unit; therefore to fit these devices a special sink is often required. When installing a waste disposal unit a tubular trap must be fitted to its outlet because the units do not have their own integral trap fitted.

When using these units it must be remembered that water must be flowing down the waste pipe, otherwise the machine and pipe would soon become blocked up. Should the machine become jammed a special tool is provided to turn the cutting head to loosen the obstruction and on many units there is a cut-off switch located on the machine; this is provided to prevent the motor from becoming burnt out should it be jammed. Upon freeing the cutting head the cut-off switch must then be pressed to reset it.

WC macerator pump

A special packaged unit consisting of a macerator and a pump which can be installed behind a WC pan to collect the discharge and macerate up any solid matter to allow it to be pumped vertically up to 4 m or horizontally up to 50 m and discharged into a small 19 mm nominal bore discharge pipe.

Macerator pumps are only permitted to be installed if there is also access to a WC discharging directly into a gravity system of drainage; the reason for this is that should there be an electrical failure the machine is put out of action. The holding tank to these machines needs to be ventilated to allow for its gravity filling and to facilitate emptying; therefore, to prevent unnecessary odours emptying into the room, it is recommended that the vent be extended and terminated externally to a safe position.

The electrical connections to any form of mechanical disposal unit must be via an unswitched fused spur outlet with the correct size fuse fitted and a neon light indicator, and not simply connected to a conventional plug and socket.

Branch discharge pipes from waste disposal units and macerators require steeper gradients than is normal for waste appliances.

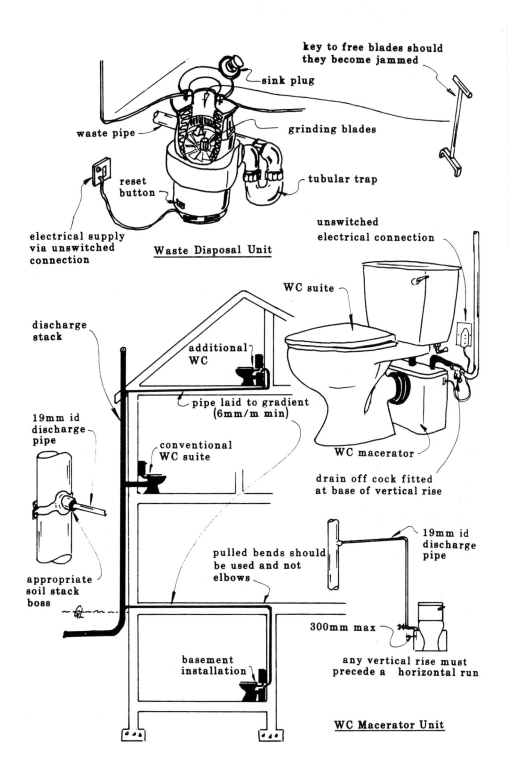

key to free blades should
they become jammed

sink plug

waste pipe

grinding blades

reset
button

tubular trap

electrical supply
via unswitched
connection

Waste Disposal Unit

unswitched
electrical connection

WC suite

discharge
stack

additional
WC

pipe laid to gradient
(6mm/m min)

19mm id
discharge
pipe

conventional
WC suite

WC macerator

drain off cock fitted
at base of vertical rise

appropriate
soil stack
boss

pulled bends should
be used and not
elbows

19mm id
discharge
pipe

300mm max

basement
installation

any vertical rise must
precede a horizontal run

WC Macerator Unit

Mechanical Disposal Units

Pipe Sizing of Sanitary Pipework

Relevant British Standard
BS 5572

The size of the branch discharge pipework has previously been identified under single stack system (page 256); however, the main vertical discharge pipe will be based upon the amount of appliances connected and type of building. The sizing of discharge stacks for commonly-used arrangements of pipework is often based on the following guide-lines, allowing for one group of appliances on each level:

☐ 100 mm diameter in buildings up to 3 levels with no additional ventilation pipes;
☐ 100 mm diameter in buildings up to 12 levels where the vertical discharge stack is ventilated with a 50 mm diameter ventilating stack (ventilated stack system);
☐ 150 mm diameter in buildings up to 24 levels, again requiring the main discharge stack to be ventilated as above.

Note: The above general guide is based on the assumption that there are no offsets in the vertical discharge stack and that the drainage system is not prone to surcharging and no intercepting trap is fitted, in which case additional ventilation may be required.

In very tall buildings, or buildings which have groups of appliances connected to the main discharge stack, the pipe size is generally found by the discharge unit method in which each appliance is given a discharge unit value to represent the average amount of water that would flow through a discharge pipe or drain. By adding all the required discharge units together the required discharge pipe size and gradient for horizontal pipes can be found from Tables 1 and 2 opposite.

Example: Find the diameter and gradient of a drain serving a school with 20 WCs, 25 wash basins, 6 urinals and 4 sinks.

$$
\begin{aligned}
\text{WCs} &= 20 \times 28 &= 560 \\
\text{basins} &= 25 \times 6 &= 150 \\
\text{urinals} &= 6 \times 0.3 &= 1.8 \\
\text{sinks} &= 4 \times 27 &= 108 \\
\hline
& & 819.8
\end{aligned}
$$

Sanitary appliance	Discharge unit values		
	Domestic use[1]	Commercial use[2]	Congested use, e.g. schools[3]
9-litre WC	7	14	28
Sink	6	14	27
Wash basin	1	3	6
Bath	7	18	18
Urinal	–	0.3	0.3
Washing machine	4	4	4

Domestic appliance used approximately every 20 mins
Commercial appliance used approximately every 10 mins
Congested appliance used approximately every 5 mins

Therefore 820 discharge units will need to be allowed for. This suggests (see Table 1) a 125 mm vertical stack. However, if the discharge were to pass through a horizontal pipe or drain, Table 2 indicates that a 100 mm diameter pipe laid to a gradient of 1:22 would be sufficient. One must remember that if the vertical stack above, discharging into this drain, is larger than 100 mm the larger size must be maintained.

Table 1 Maximum number of discharge units for vertical stacks

Internal bore (mm)	Maximum discharge units
50	10
65	60
75*	200
90	350
100	750
125	2500
150	5500

*Not more than one siphonic WC with a 75 mm outlet

Table 2 Maximum number of discharge units for horizontal pipes or drains

Internal bore (mm)	Gradient		
	1 in 111	1 in 45	1 in 22
32	–	1	1
40	–	2	8
50	–	10	26
65	–	35	95
75	–	100	230
90	120	230	460
100	230	430	1050
125	780	1500	3000
150	2000	3500	7500

(Tables 1 and 2 reproduced with permission from BS 5572 *Code of practice for sanitary pipework.*)

7 Sanitation

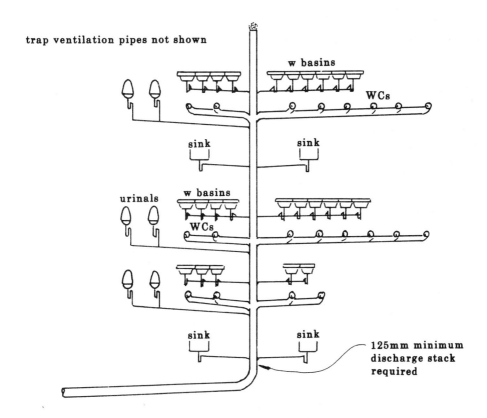

trap ventilation pipes not shown

w basins

WCs

sink   sink

urinals   w basins

WCs

sink   sink

125mm minimum discharge stack required

Pipe Sizing of Sanitary Pipework

Testing of Sanitary Pipework

Relevant British Standard
BS 5572

Generally all inspection and testing should be made during the installation stage, especially where the pipework will be inaccessible upon completion. Two tests are carried out to sanitary pipework, these being identified as follows:

Soundness test (air test) A test to see if the system is airtight. The procedure is to insert testing bags or drain plugs into any open ends and fill all the appliance traps with water. It is also advisable to allow some water to cover the test plugs in order to provide a suitable seal. A rubber hose connecting a manometer and hand bellows is connected to one drain plug. By operation of the hand bellows air is pumped into the system to give a water head pressure of 38 mm, as indicated on the manometer. A plug cock on the air inlet tube is now closed and the pressure should hold for 3 min with no pressure drop.

Should a leak be evident it will need to be found by maintaining a small air pressure within the system and applying leak detection fluid, such as washing up liquid and water, to the joints; any leakage will be indicated by the formation of bubbles. The system may be charged with smoke although this is not recommended for plastic pipework or where rubber components may be adversely affected.

Performance test A test which is carried out to above-ground sanitary pipework to ensure that a minimum of 25 mm of water trap seal is retained in every trap when the pipework is subjected to its worst possible working conditions. Each of the following tests should be carried out a minimum of three times and before each test the trap should be recharged. The depth of the water seal should be measured with a dipstick (see figure).

Tests for siphonage in branch discharge pipes The test for self-siphonage is carried out by filling the sanitary appliance to overflowing level and removing the plug. To test for induced siphonage several appliances should be discharged together and all the traps on the pipe run being tested should be measured. The worst conditions occur when the sanitary appliances furthest away from the drain or discharge stack are discharged.

Tests for siphonage and compression in discharge stacks A selection of sanitary appliances as indicated in the table should be discharged at the same time from the highest floor(s) thus giving the worst pressure conditions.

Example: A seven-storey block of flats has two sanitary appliances of each type on each floor, i.e. two baths, two sinks, two wash basins and two WCs. The amount and kind of appliances to be discharged to give a satisfactory test will therefore be:

$$2 \times 7 = 14 \text{ of each kind of appliance}$$

Thus from the table we see that one WC, two sinks and one wash basin should be simultaneously discharged from the top floor.

Performance testing in discharge stacks

Type of building	No. of each type of sanitary appliance on stack	No. of appliances to be discharged together		
		9 litre WC	sink	basin
Domestic	1–9	1	1	1
	10–24	1	2	1
	25–35	1	3	2
	36–50	2	3	2
	51–65	2	4	2
Commercial and public	1–9	1–2	–	1–2
	10–18	1–2	–	2–3
	19–26	2–3	–	2–3
	27–50	2–3	–	3–4
	51–78	3–4	–	4
	79–100	3–4	–	5

(Reproduced with permission from BS 5572 *Code of practice for sanitary pipework.*)
Note: baths, showers, urinals and spray taps are ignored because their discharge does not adversely affect the normal peak flow load. In public buildings such as schools, where very congested periods can be expected, the larger figure should be chosen.

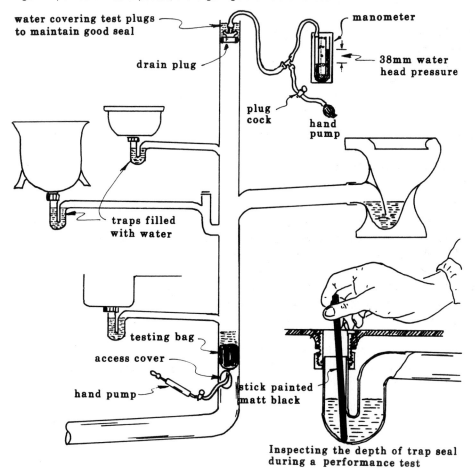

Inspecting the depth of trap seal during a performance test

Testing of Sanitary Pipework

Maintenance and Periodic Inspection

Relevant British Standard
BS 5572

Any drainage system should be kept in a clean, sound condition in order to maintain its maximum efficiency. During periodic inspections access covers will need to be checked for operation; where cast iron or steel ventilating stacks have been employed a flush through with water or rodding may be required to remove rust accumulations at offsets and bends.

Occasionally discharge pipes and drains become blocked due to an assortment of causes and it is for the plumber to overcome the problem of unblocking the pipe with the least inconvenience. The methods used can vary depending upon circumstances from using drain rods to a suction plunger, or even the use of chemicals may be needed.

Manual cleaning

Drain rods are passed down the pipe until the obstruction is met; then, by giving a few blows, the blockage is dislodged and with the pressure caused by the build up of water it is often washed away.

A variation of this is to pass a spring or drain auger down the pipe; this is rotated, thus dislodging the blockage; the spring could either be hand-held or machine-operated.

Possibly one of the most frequently employed methods of removing blockages is to use a plunger; the reason why the plunger proves so successful is because it creates a pressure on the blockage of at least that of atmospheric pressure (101.3 kN/m^2). As a plunger is withdrawn it leaves a void behind; thus a partial vacuum is formed. When plunging sink wastes it is essential that the overflow pipe, if connected to the waste, is blocked up otherwise air will travel down this and relieve the negative pressure within the pipe.

Chemical cleaning

When carrying out any work of this nature protective clothing and eye shields should be worn and all work upon completion should be thoroughly flushed and washed down.

A descaling fluid containing 15–30% inhibited hydrochloric acid and 20–40% phosphoric acid is poured into the pipes in small quantities at predetermined points, or it can be applied via a drip feed into the pipe at a rate of about 4 litres every 20 min. When carrying out any descaling all windows should be opened to ensure good ventilation. Most acid descaling fluid will attack linseed oil bound putty; therefore prolonged contact with these jointing materials should be avoided.

Should the problem be less severe and the discharge pipework lined or blocked only with grease or soap residues, the pipework should be flushed with very hot water which has 1 kg of soda crystals dissolved in every 9 litres. **Note**: soda crystals should not be confused with caustic soda.

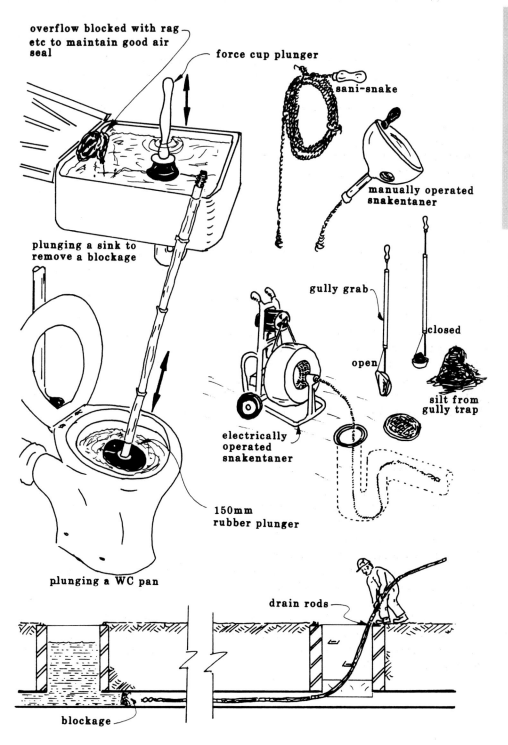

overflow blocked with rag etc to maintain good air seal

force cup plunger

sani-snake

plunging a sink to remove a blockage

manually operated snakentaner

gully grab

closed

open

silt from gully trap

electrically operated snakentaner

150mm rubber plunger

plunging a WC pan

drain rods

blockage

Maintenance and Periodic Inspection

Part 8

Drainage

Below Ground Drainage

Relevant British Standard
BS 8301

All drainage work must be carried out in accordance with Part H of the Building Regulations. Drainage is primarily divided into two groups: surface and foul water systems. *Surface water* is the water from roofs and the surrounding ground, whereas *foul water* is water which is contaminated by soil, waste and trade effluent. Clearly surface water does not need to be treated and may discharge into a local water course. Any drainage system must convey all surface water or liquid sewage away from the building in the most speedy and efficient way, possibly to the sewer or other discharge point without risk of nuisance or danger to health and safety. When designing any system one must observe the following principles:

☐ Provide adequate access points;
☐ Keep pipework as straight as possible between access points, and for all bends over 45° an access point should be provided;
☐ Ensure all pipework is adequately supported;
☐ Ensure the pipe is laid to a self-cleansing gradient;
☐ The whole system must be watertight, including inspection covers;
☐ Drains should not run under a building, unless this is unavoidable or in so doing they would considerably shorten the route of pipework.

There are three designs of below ground drainage:

Combined system A system in which one pipe is used to convey foul and surface water. When this system is used all points in the system open to the atmosphere must be trapped; the only exception to this would be ventilating pipes and fresh air inlets. The advantage of this system over separate and partially separate systems is that it is cheaper and easier to install; also, during periods of rain the whole system gets a good flush through. The disadvantages of this system are also ones of cost, because all water (foul and surface) has to be treated at the sewerage works. Also, at times of heavy rainfall inadequately sized drains could be prone to surcharging, which makes water course pollution very probable.

Separate system In this system the foul water is conveyed in one pipe to the sewage treatment works, and all the surface water is conveyed in another pipe completely independently of the foul water; thus there is no need for water treatment. Connections to surface water drains do not need to be trapped because there should be no unhealthy smells, but with the foul water drain all connections must be trapped. One problem with this system is the danger of cross-connections, i.e. foul water being connected to a surface water drain.

Partially separate system A compromise between the combined and separate systems. It consists of two pipes, one for surface water and one for foul water; plus a limited amount of surface water. The system allows one the opportunity to overcome particular design problems such as those which may occur where no local watercourse is available and a soakaway may prove ineffective.

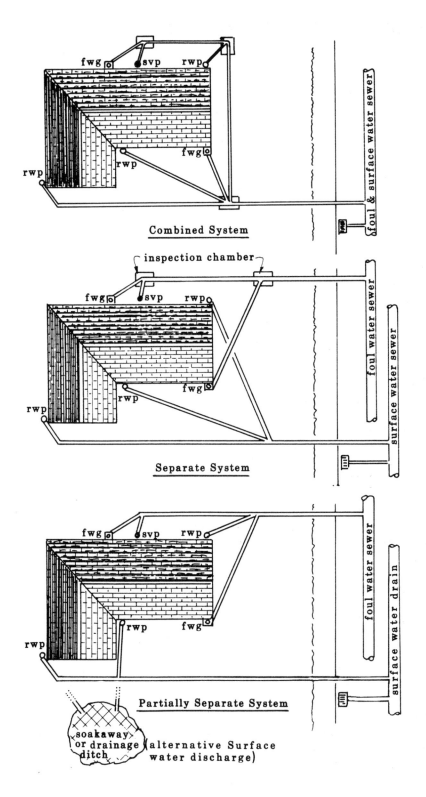

8 Drainage

Combined System

inspection chamber

foul & surface water sewer

Separate System

foul water sewer

surface water sewer

Partially Separate System

foul water sewer

surface water drain

soakaway or drainage ditch (alternative Surface water discharge)

fwg   svp   rwp

rwp   rwp   fwg

Below Ground Drainage

Protection of Pipework

General provisions

When the drainage pipe is positioned in the ground it must be pro due to ground movement, etc. This is usually achieved by laying the pipe on a granular bedding material and covering the pipe with soil which is free from large stones or other such material. Different circumstances, such as the size of the pipe and the depth to which it is laid, call for different types of bedding. The bedding can either be for rigid or flexible pipes.

Rigid pipe materials include clayware, concrete and cast iron whereas flexible pipes include the various plastics. If the ground is stable some authorities will permit drainage pipes to be laid directly onto the trench bottom, although the main difficulty is ensuring a steady gradient. The purpose of the granular material is to distribute any excessive loads more evenly around the surface of the pipe, preventing its distortion and possible damage.

Where a pipe does not have the recommended cover it may require additional protection from damage by several methods including encasement in concrete in the case of rigid pipes, allowing for movement at joints or covering the bedding material with some form of paving slab.

Allowance for settlement protection of buildings

Drains below buildings Where a drain is run under a building it should be surrounded with at least 100 mm of granular material. If the crown (top) of the pipe is within 300 mm of the underside of the oversite concrete slab it should be encased in concrete and be incorporated into the slab.

Drains penetrating walls Should a drainage pipe have to run through a wall or foundation, special precautions will need to be taken to ensure the pipe does not fracture. This is best achieved by either of the following methods:

(1) Forming an opening through the wall giving a 50 mm space all around the pipe which is masked off with a rigid sheet material.
(2) Bedding into the wall a short length of pipe onto which is connected two 600 mm long pipes, either side of the wall (all joints being made good using flexible connectors). Should the pipe or wall move, the three pipes would act as rockers, allowing for movement.

Trenches close to building foundations Where a drainage trench is excavated lower than the foundations of any building, and within 1 m, the trench should be filled with concrete up to the lowest level of the foundation. Distances greater than 1 m should be filled with concrete to a level equal to a distance from the building, less 150 mm (see the Building Regulations).

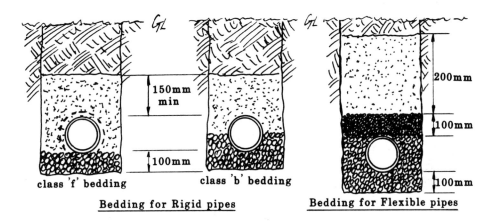

150mm min

100mm

200mm

100mm

100mm

class 'f' bedding

class 'b' bedding

Bedding for Rigid pipes

Bedding for Flexible pipes

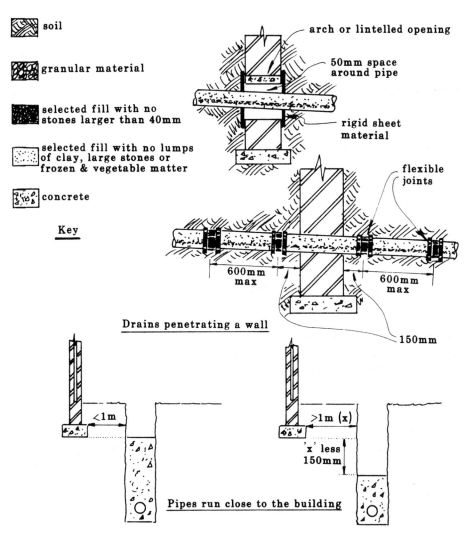

soil

granular material

selected fill with no stones larger than 40mm

selected fill with no lumps of clay, large stones or frozen & vegetable matter

concrete

Key

arch or lintelled opening

50mm space around pipe

rigid sheet material

flexible joints

600mm max

600mm max

Drains penetrating a wall

150mm

<1m

>1m (x)

'x' less 150mm

Pipes run close to the building

Protection of Pipework

Gullies and Traps

The design of a drainage system may allow the incorporation of any one of a whole series of gullies or traps to help ensure that design performance is met. In general the gully is a drainage fitting designed to receive surface and/or foul water from waste pipes. The purpose of a gully is to provide a trap preventing odours from the drain entering the atmosphere. All new works must discharge the water from appliances below the cover or grating and in many cases a back or side inlet gully is used to assist in this design. The old method of discharging the water into a chamber above the grating is no longer permissible. Trapless gullies are also available but should only be used for surface water drains.

Yard and garage gullies A large gully which includes a sediment bucket to collect any grit or silt which might otherwise block the outlet pipe. The bucket should periodically be removed for cleansing purposes.

Anti-flood gullies and valves A fitting designed to prevent surcharging of surface water drains. With the ball type shown, when backflow or surcharging occurs the ball rises onto the underside of the access cover until it finally sits into the rubber seating, preventing any further flow in either direction until the flood water subsides. With the trunk valve any backflow causes the float to rise, thus turning the flap valve to the closed position.

Grease trap and converters A grease trap is a device which houses a quantity of water and is designed so that should grease from canteen kitchens be discharged into the drains it would, when reaching the water, cool and solidify. Lighter solids would rise to the surface and collect in a solid cake; heavier solids would sink and collect in the galvanised tray. Periodically the tray should be lifted out, removing all the grease which has accumulated.

Some grease traps are designed to be fitted under sinks but in most cases the trap is fitted externally to the building. The grease converter allows the effluent to pass over a series of baffles which cause the grease to form in globules on the surface. Applying a regular dosing of a mixture of micro-organisms, enzymes and food supplements, in powder or liquid form, to the globules of grease converts it into a water-soluble biodegradable product.

Intercepting trap A type of running trap usually found in older properties in the inspection chamber nearest the main sewer – sometimes called a disconnecting trap because it disconnects the drain from the sewer. The prime function of an intercepting trap is to prevent sewer gas from entering the house drains, but nowadays because drainage systems are better designed and sewer gases seldom cause trouble in most cases an intercepting trap is omitted. In the past the intercepting trap proved a common cause of blockages owing to the collection of debris in the trap.

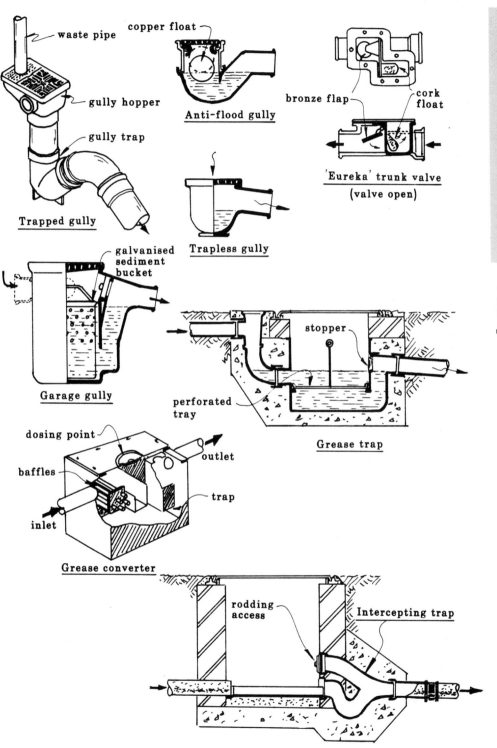

- waste pipe
- copper float
- gully hopper
- gully trap
- bronze flap
- cork float

Trapped gully

Anti-flood gully

'Eureka' trunk valve (valve open)

Trapless gully

- galvanised sediment bucket

Garage gully

- stopper
- perforated tray

Grease trap

- dosing point
- outlet
- baffles
- trap
- inlet

Grease converter

- rodding access
- Intercepting trap

Gullies and Traps

Provision for Access

Relevant British Standard
BS 8301

To enable internal inspection and testing of a drainage system or to provide a route in which the clearance of blockages can be achieved it is essential that sufficient provision is made for the internal access of the pipe. An access point should be provided at the following locations:

- ☐ At or near the head of a drainage run.
- ☐ At changes of gradient or direction (i.e. bends).
- ☐ At junctions or branches, unless it is possible to clear blockages from another access point.
- ☐ At changes in pipe diameter.
- ☐ Between long drainage runs. This will be dependent on the type of access provided (see the Building Regulations).

Three types of access will be found:

Rodding eyes A capped extension on the pipe where access can be gained to a drain or any discharge pipe for the purpose of cleaning with rods or inspection. It is possible to design a 'closed' system of drainage in order to reduce the number of inspection chambers found in the more traditional systems. The rodding point system, as it is generally known, is often run in uPVC pipe and with the omission of inspection chambers reduces the cost of installation considerably.

Access fittings A fitting, such as a bend, branch or gully, which has a cover fitted, usually bolted to the fitting, in order to gain access. The cover may be located above ground or at ground level such as in the case of a gully. It may be located below ground, in which case it will need to be incorporated into an inspection chamber or manhole; alternatively a raising piece could be incorporated to allow termination at ground level. The access fitting, unlike the rodding eye, allows rodding in more than one direction.

Inspection chambers and manholes A chamber constructed of brick, concrete or plastic and designed to expose a section of open pipe, in the form of a channel at its base. The definition of an inspection chamber or manhole is based on size. If the chamber is large enough the work in it is identified as a manhole, which would certainly be the case in all chambers over 1 m deep. All chambers are provided with some form of cover located at ground level and when positioned internally within a building it should be bolted down with a greased double seal incorporated to prevent the passage of odours.

When a chamber is constructed in brickwork or concrete its wall thickness should be adequate to resist any external pressures caused by the surrounding ground; in all cases it should be at least 200 mm thick. The base of the chamber should be benched up to allow any rising water flow back down into the channel. In chambers over 1 m deep step irons or a ladder will need to be included.

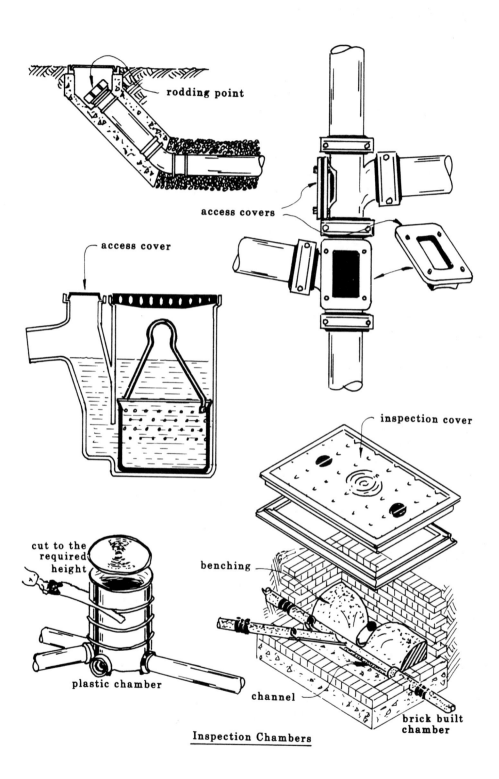

rodding point

access covers

access cover

inspection cover

cut to the
required
height

benching

plastic chamber

channel

brick built
chamber

Inspection Chambers

Provision for Access

Connections to Existing Systems

Relevant British Standard
BS 8301

When joining into an existing drain run several methods can be adopted and in most cases the most suitable method is to connect into an existing inspection chamber or construct a new one. Sometimes a direct connection has to be made, for example where a house drain is to join a public sewer.

The method of joining into the pipe would depend upon the material but in many cases either a junction block or saddle junction could be used, which must be connected into the top half of the drain. If the pipe into which you are cutting is less than 225 mm in diameter it is best to insert a new branch junction because a hole cut in such a small pipe would weaken it.

When a drainage pipe at high level is to join a drain at a lower level, where possible the drain should be run at a gradient adjoining the two. However, if the difference in heights between the two drains is excessive or space prevents the gradient being self-cleansing, access should be afforded to the pipe for future cleansing purposes. Either of two methods can be adopted for this: the ramp or the backdrop.

The ramp This may be used where the difference in levels between the two drains does not exceed 680 mm. It consists of an open channel formed in the benching of an inspection chamber to join the two drainage pipes.

The backdrop or tumbling bay This consists of a section of vertical drainage pipe joining the invert level in an inspection chamber and a drain pipe at higher level. A backdrop would be used if the vertical distance between invert levels of the drain exceeded 680 mm. The backdrop may be installed inside the inspection chamber if it is run in cast iron or plastic, but should clayware pipe be used it must be fitted externally to the chamber and encased in concrete.

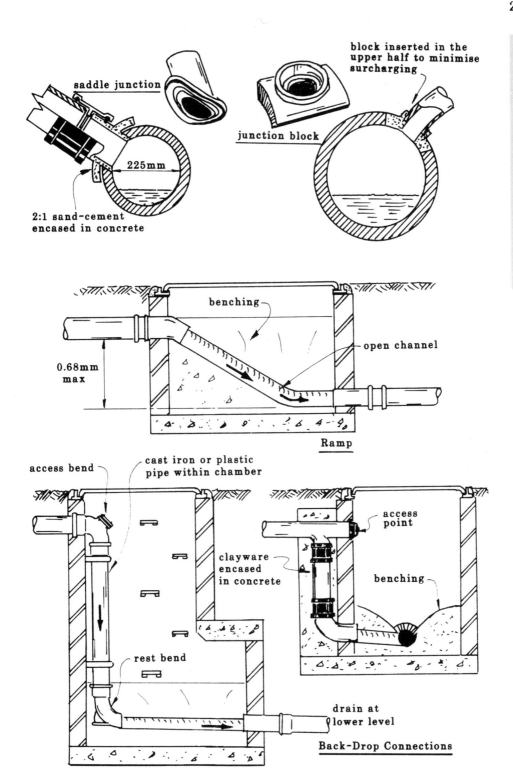

block inserted in the
upper half to minimise
surcharging

saddle junction

junction block

225mm

2:1 sand-cement
encased in concrete

benching

open channel

0.68mm
max

Ramp

access bend

cast iron or plastic
pipe within chamber

access
point

clayware
encased
in concrete

benching

rest bend

drain at
lower level

Back-Drop Connections

Connections to Existing Systems

Determining Drainage Levels

Relevant British Standard
BS 8301

Choice of gradient

The flow capacity of a pipe will depend upon the gradient, fall or incline on which the pipe is laid. Increasing the gradient increases the capacity. It is essential that the pipe is laid to a self-cleansing gradient with a flow velocity of between 0.75 m/s and 1.54 m/s. If the water flows too slowly it will have insufficient velocity and impetus to carry with it any solid matter; however, where it flows too fast the water will leave the solids behind.

The amount of flow passing through the pipe also has a bearing upon the gradient; and where intermittent flow can be expected, giving shallow depths of water flow, it is advisable to increase the gradient. Maguire's Rule is sometimes used to determine self-cleansing gradients with intermittent or low-flow rates and is based on a calculation in which the pipe diameter is divided by 2.5: for example, 100 mm ÷ 2.5 = 40, therefore 1:40 (see table).

Where continuous flows can be expected the gradient for 100 mm pipes can be reduced to 1:80, providing one WC is connected; and for 150 mm pipes a gradient no flatter than 1:150 may be permitted, providing at least five WCs are connected to the drainage run.

Maguire's Rule

Pipe diameter (mm)	Recommended fall
100	1 in 40
150	1 in 60
225	1 in 90

Setting out the fall

The method used to ensure the pipe is laid to the correct gradient will depend upon the length of drainage run, and for short sections a gradient board, or incidence board, is used. The gradient board is a plank of wood cut to the required gradient of a drain and used in conjunction with a spirit level. If a drain run is to have a fall of 1 in 40 the drop in depth over 1 m would be (1 ÷ 40 = 0.025 m) 25 mm. So to make a simple gradient board, take a straight plank 1 m long and cut it at an angle (see figure).

Where a long drainage run is to be installed the method used to determine the fall should be carried out using sight rails and traveller (sometimes called a boning rod). First the site rails are positioned at either end of a drainage run at different heights, the difference being that of the required gradient.

This process is done by a site surveyor, using a site level, or by using a water level to give an accurate transference of levels between each end of the drainage run. Then the trench is excavated to the required fall or backfilled as necessary. Two operatives are then used, one sighting his eye between the site rails and instructing his partner to raise or lower the traveller by lifting or lowering the pipe as necessary (see figure).

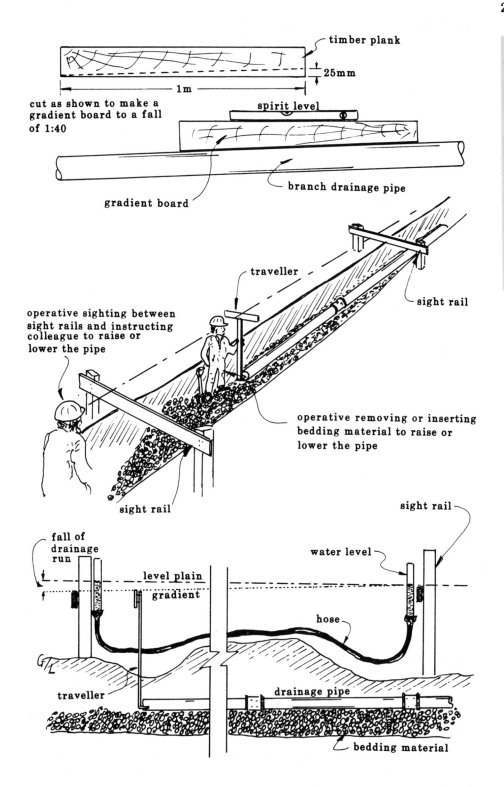

timber plank

25mm

1m

cut as shown to make a
gradient board to a fall
of 1:40

spirit level

branch drainage pipe

gradient board

traveller

sight rail

operative sighting between
sight rails and instructing
colleague to raise or
lower the pipe

operative removing or inserting
bedding material to raise or
lower the pipe

sight rail

sight rail

fall of
drainage
run

water level

level plain

gradient

hose

traveller

drainage pipe

bedding material

8 Drainage

Determining Drainage Levels

Pipe Sizing of Drainage Pipework

Relevant British Standard
BS 8301

The minimum diameter of any drainage pipe is 75 mm; however, if the pipe is to convey soil, water or trade effluent the minimum size is 100 mm. In determining the pipe size for a drainage system one must first calculate the probable expected flow rate of discharge water in litres per second.

In *soil and waste drainage systems* the flow rate is based upon the number of appliances connected to the system and its likely discharge, bearing in mind that not all appliances will be in use at the same time (see page 264, Pipe Sizing of Sanitary Pipework). For the typical domestic dwelling Table 1 can be used.

Table 1 Approximate flow rate from a typical domestic household with minimal sanitary accommodation

No of dwellings	Flow rate litre/s
1	2.5
5	3.5
10	4.1
15	4.6
20	5.1
25	5.4
30	5.8

In *surface water drainage systems* the drain should be large enough to carry the maximum flow, from the paved and other surfaces, with an allowance for that which evaporates or soaks into the surrounding ground. This is known as an impermeable factor and will be as indicated in Table 2.

Table 2

Surface	Impermeable factor
Roofs	0.8–0.95
Asphalt pavements	0.8–0.9
Jointed stone pavements	0.8–0.85

In areas where water drains readily into the ground the lower number is chosen.

In the UK a rainfall intensity of 0.05 m per hour (50 mm/h) should be assumed and is used in the following calculation to find the volume of flow in litres/s:

$$\frac{\text{Area (m}^2) \times \text{rainfall intensity} \times \text{impermeable factor}}{\text{seconds in one hour}} \times 1000 = \text{litres/s}$$

Example: The volume of flow in litres/s from a roof measuring 72 m^2 would be:

$$\frac{72 \text{ m}^2 \times 0.05 \text{ m/h} \times 0.9}{3600} \times 1000 = 0.9 \text{ litres/s}$$

It is only in combined systems of drainage that both foul and surface water discharge are added together; otherwise they are simply two individual systems, each having its own requirements.

If we take as an example a group of five houses which are to connect to a combined system of drainage in which the total roof area measures 375 m², and the surrounding paths are to be 400 m², the probable flow to allow for (in litres/s) would be:

(1) Flow from sanitary accommodation (from Table 1 opposite) = 3.5
(2) Flow from roofs 375 × 0.05 × 0.95 ÷ 3600 × 1000   = 4.9 +
(3) Flow from paths 400 × 0.05 × 0.80 ÷ 3600 × 1000   = 4.4
                                                      12.8 litres/s

Having determined the flow rate, we can now refer to the graph below to find the minimum required size of pipe, bearing in mind the gradient at which the pipe is to be laid (see page 282). Thus, in this example, for a combined system serving five houses with a flow rate of approximately 13 litres/s, and with a gradient of between 1:60 and 1:150, a pipe size greater than 100 mm will be required: 150 mm pipe would therefore be chosen.

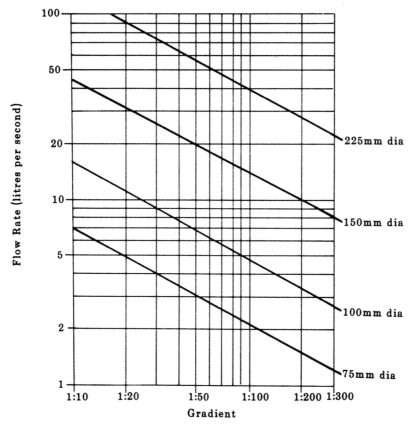

Discharge capacities of drains flowing $\frac{2}{3}$ proportional depth

Pipe Sizing of Drainage Pipework

(Reproduced with permission from BS 8301 *Code of practice for building drainage.*)

8 Drainage

Eaves Guttering

Relevant British Standard
BS 4576

The guttering fixed at the lower edge of a pitched roof. There are many designs of eaves gutter, the most common being half round, square line and 'O gee'. Probably the most common material used today for guttering is plastic, although it is also made in cast iron and galvanised pressed steel. The method of securing eaves guttering to the building depends upon the building design but in general fascia brackets are used.

When installing any guttering material allowances for expansion must be taken into account, especially with plastic guttering materials as the finished gutter will be exposed to constant temperature changes. Occasionally putty is used to join the non-plastic materials, but this is not to be recommended because it dries hard and there is no allowance for thermal movement; instead Plumbers Mait or silicone rubber should be used, these being non-setting.

The fall or gradient of a gutter should be approximately 1 in 600. This fall would scarcely be noticeable to the human eye; for example, over 10 m the fall would only be $(10 \div 600 = 0.017)$ 17 mm. Once the fall has been determined the brackets should be fixed at about 1 m intervals, as shown. Once the gutter has been completed it is connected to the drain via a rainwater pipe.

Gutter size The size of the gutter is dependent upon the size of the roof area to be drained. The effective roof area (in m^2) for pitched roofs is different to that for flat roofs and needs to be found by calculation. By observing Table 1 it will be found that for pitched roofs a greater 'run-off' or flow will be experienced and must be allowed for.

Table 1   Calculation of roof area

Roof pitch	Effective area (m^2)
Flat	Plan area
30°	Plan area × 1.15
45°	Plan area × 1.4
60°	Plan area × 2.0
70° to vertical	Elevation area × 0.5

Example: If the plan area of a 45° pitched roof measures 6 m × 4 m, the effective area will be:

$$6 \times 4 \times 1.4 = 33.6 \ m^2$$

Once the effective area has been determined the gutter size will be found by referring to Table 2, or where different guttering materials are used, the manufacturer's data. Continuing with the above example, in which the effective area was 33.6 m^2, a half round gutter size of 100 mm minimum can be chosen.

Table 2 Half round gutter size (outlet at one end)

Max effective area (m^2)	Gutter size (mm)	Outlet size (mm)	Flow capacity (litres/s)
18.0	75	50	0.38
37.0	100	63	0.78
53.0	115	63	1.11
65.0	125	75	1.37
103.0	150	89	2.16

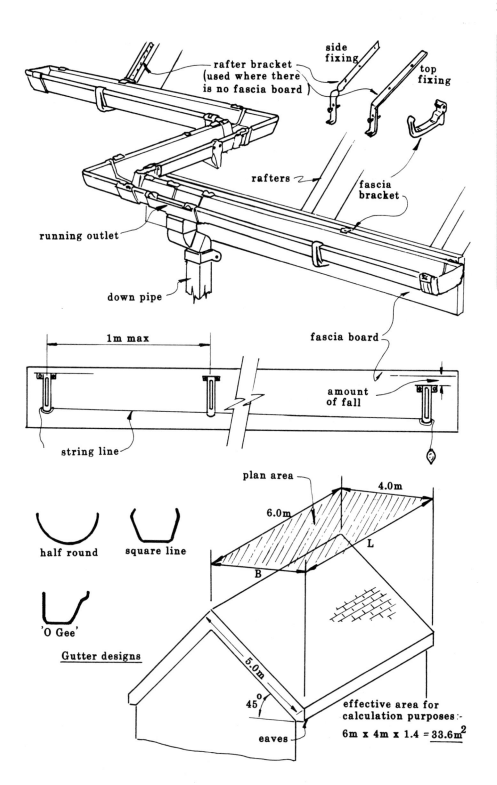

side fixing

rafter bracket
(used where there
is no fascia board)

top fixing

rafters

fascia bracket

running outlet

fascia board

down pipe

1m max

amount of fall

string line

plan area

4.0m

6.0m

L

B

half round

square line

5.0m

'O Gee'

Gutter designs

45°

eaves

effective area for
calculation purposes:-
6m x 4m x 1.4 = $\underline{33.6\text{m}^2}$

Eaves Guttering

Rainwater Pipes

Relevant British Standard
BS 4576

The pipe used to convey the surface water collected from the roof to the drain at ground level. In most cases this pipe is run externally down the outside of the building in which, normally, no jointing medium is used but the spigot simply enters the socket.

If the rainwater pipe is run internally the joints must be made watertight. As with eaves guttering, it is essential to allow for the expansion of the material – otherwise buckling or cracking of the pipes will result. For the sake of appearance, it is essential to ensure that the rainwater pipes are fixed perfectly vertical and to assist in this operation it is best to use a plumb bob.

The termination of the rainwater pipe at the lower level could be either by a rainwater shoe, discharging into a gully, or preferably the pipe is run directly into a back inlet gully.

The size of the rainwater pipe should be at least that of the gutter outlet, and where a pipe serves more than one gutter it should be as large as the combined areas of both outlets.

Rainwater connections to discharge stacks

In some areas with a combined system of drainage the local authority will permit the foul and waste water discharge stacks to receive rainwater from roofs. Designing a system in this way can save on the cost of installing separate rainwater pipes.

To avoid excessive air pressure fluctuations within the discharge stack this method of rainwater disposal is not to be recommended for buildings over ten storeys in height, or for the removal of rainwater from roof areas exceeding 40 m^2 per stack. The main disadvantage of designing above-ground drainage in this way is the problem of flooding if a blockage occurs in the discharge stack. The rainwater pipe should be connected to the stack via a branch connection (see figure); this ensures a free unrestricted flow of air through the ventilating pipe.

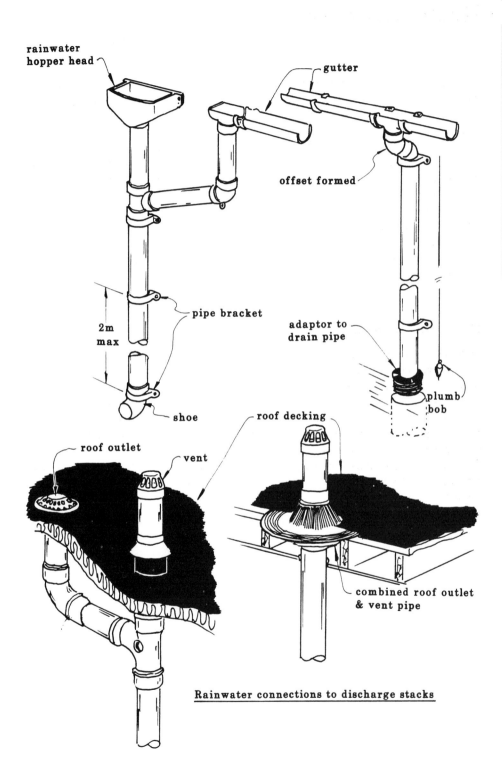

rainwater
hopper head

gutter

offset formed

pipe bracket

2m
max

adaptor to
drain pipe

plumb
bob

shoe

roof outlet

vent

roof decking

combined roof outlet
& vent pipe

Rainwater connections to discharge stacks

Rainwater Pipes

Soakaways, Cesspools and Septic Tanks

Relevant British Standard
BS 8301

Soakaway

This is a hole, sited well away from the building, which is generally filled with bricks, rubble or stones to prevent the side walls caving in. The purpose of a soakaway is to receive surface water and it is often used in country areas where there is a lack of sufficient surface water drainage pipes. Soakaways should only be used in areas which will allow the water to percolate into the surrounding ground.

The size and distance from the building is usually specified by the local authority, which must be notified of its use, but in general a soakaway should not be sited nearer the building than 5 m and its size should be equal to 12 mm of rainfall over the area to be drained. For example, the minimum size of a soakaway for an area of 125 m^2 should be 125 × 0.012 m = 1.5 m^3, measured below the drain connection invert level. If possible, the soakaway should be sited on ground which slopes away from the building in case of flooding.

Cesspool (cesspit)

A watertight container used under ground for the collection and storage of foul water and crude sewage. A cesspool might be used in areas where the main drainage has not been connected to the property and local sewage treatment such as a septic tank is impracticable. The contents from a cesspool will need to be removed regularly (and before completely full) for proper disposal. Where plastic cesspools are used these will need to be anchored down with concrete to prevent a high water table causing them to be lifted out of the ground.

Septic tank

A sewage disposal unit sometimes used in rural areas. The house drain is connected to a septic tank with a natural drainage irrigation trench or subsoil drain through which water can drain into the soil. Enough trenching must be constructed to ensure that flooding of the ground surface does not occur because the effluent from septic tanks contains pathogenic bacteria (diseased bacteria); it also must not be allowed to flow directly into streams or underground rivers, etc.

The septic tank usually consists of a double compartment tank which can be constructed of concrete, but more commonly nowadays they are made of reinforced glass fibre. As the liquid sewage flows into the tank, the solids settle out and most are decomposed by anaerobic bacterial action, no chemicals being required. Periodically the accumulation of sludge and remaining solids must be removed by cleansing contractors.

Before the installation of a septic tank it will be necessary to obtain local authority approval and in most cases the siting must be a minimum distance from habitable buildings of at least 15 m, and preferably sloping away from the building.

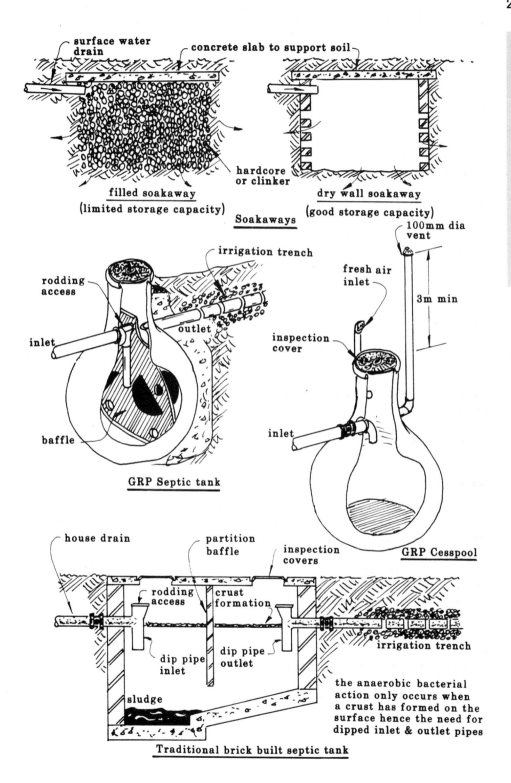

surface water drain — concrete slab to support soil

hardcore or clinker

filled soakaway
(limited storage capacity)

dry wall soakaway
(good storage capacity)

Soakaways

irrigation trench

rodding access

inlet

outlet

baffle

GRP Septic tank

100mm dia vent

fresh air inlet

3m min

inspection cover

inlet

GRP Cesspool

house drain

partition baffle

inspection covers

rodding access

crust formation

dip pipe inlet

dip pipe outlet

irrigation trench

sludge

the anaerobic bacterial action only occurs when a crust has formed on the surface hence the need for dipped inlet & outlet pipes

Traditional brick built septic tank

Soakaways, Cesspools and Septic Tanks

Soundness Testing of Drainage Systems

Relevant British Standard
BS 8301

Generally drainage works are tested in two stages: initially as the pipework is installed, with the pipe exposed; and secondly upon completion prior to handing over. The testing of pipes is carried out separately from manholes, etc.

Prior to soundness testing A ball, 13 mm smaller in diameter than the pipe, should be passed down the pipe, between access points and accessible branch drains, to check for correct directional gradient and obvious obstructions, the ball being inserted at the highest end. It is possible to check for defects in a straight drainage run by an internal inspection, using a mirror and torch, as shown.

Water test (hydraulic test) To carry out a water test the testing equipment is set up (see figure) to give a minimum test head of 1.5 m. Note that the maximum head should not exceed 4 m at any point, which may mean carrying out the test in sections. When filling the pipe with water, it is essential to make sure there are no pockets of trapped air. When the 1.5 m head pressure has been achieved the water should be allowed to stand for 2 h, topping up as necessary. After the 2 h have elapsed a note should be kept as to the amount of water needed to maintain the test head over the next 30 min.

The pipe can be deemed sound providing the water loss does not exceed that indicated in the table.

Acceptable water loss over 30 min test period

Pipe size (mm)	litres/m run
100	0.05
150	0.08
225	0.12
300	0.15

Example: In testing a 20 m run of 100 mm pipe the water loss over the actual 30 min test period should not exceed $20 \times 0.05 = 1$ litre.

Air test (pneumatic test) The air test should be carried out (see figure), stopping all open ends with drain plugs or bags and applying an air pressure to the pipeline by blowing into the tube using a hand pump. The manometer and supply air, if possible, should be at separate locations; this ensures a positive test result over the whole system. The test pressure should be 100 mm water gauge; however, where trapped gullies and/or ground floor appliances are connected to the drain run the pressure can only be 50 mm.

Five minutes is allowed for temperature stabilisation and the air pressure readjusted. The head of water in the manometer should not fall more than 25 mm over a further 5 min test period. Where a 50 mm test was used this head should not exceed 12 mm over the test period.

Testing inspection chambers and manholes These are generally carried out by fitting a bag stopper in the outlet, to which the air can be removed at ground level and the remaining connections stopped up. All plugs must be secured to ensure they are not lost down the pipe. The chamber is filled with water and left to stand for 8 h, topping up as necessary. The acceptable criterion is that no water should be observed issuing from the outside face of the chamber.

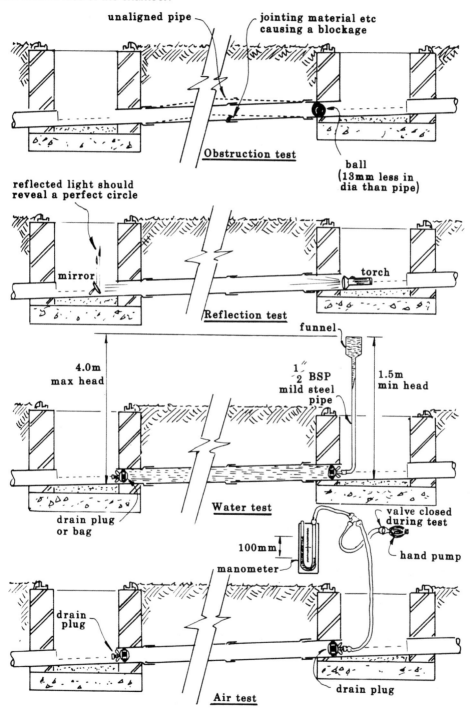

Soundness Testing of Drainage Systems

Part 9
Sheet Weathering

Lead Sheet

Relevant British Standard
BS 1178

Sheet lead is available in almost any size, although the standard width of a roll is 2.4 m, lengths being up to 12 m. Sheets of this size are very heavy to handle; therefore, in most cases, it is worth the extra charge of buying it in smaller (strip) widths varying from 150 to 600 mm. Sheet lead is available in a range of thicknesses and is colour-coded accordingly.

Lead is highly malleable and lead bossing can be carried out with ease. Due to its low tensile strength and lack of elasticity lead will creep; therefore correct installation of the material is essential. It is important to adhere to the installation requirements given in the following pages.

Lead can very easily be welded together (see page 52). The thickness of the sheet used will depend upon its location and the type of building (e.g. historic or in a position of extreme exposure). Generally for small flat roofs and most flashing applications codes 4 and 5 will prove adequate. Usually, where sheet is to be welded, code 4 is used, and if bossed code 5 is chosen. In all cases no flashing should exceed 1.5 m in length. Code 3 will be used for soakers and its maximum length must not exceed 1 m.

British Standard thickness of sheet lead

BS 1178 Code No	Thickness (mm)	Weight (kg/m²)	Colour code
3	1.32	14.97	Green
4	1.80	20.41	Blue
5	2.24	25.40	Red
6	2.65	30.05	Black
7	3.15	35.72	White
8	3.55	40.26	Orange

When new lead is first laid it will produce an initial white carbonate on its surface which will eventually be washed away by the rain but may result in the staining of brick-work, etc. To avoid this staining and provide a shiny appearance when completing the installation of any work, it is wise to apply a coating of patination oil with a cloth to the surface and lower edges or between and below laps and clips.

Lead used on a large expanse will need to have an underlay to assist in allowing for thermal movement and to isolate the sheet material from the roof structure, thus preventing electrolytic corrosion between any steel roof fixing and the lead sheet. An underlay also acts as an insulator against heat transference and sound.

Traditionally impregnated inodorous felts have been used and proved satisfactory on concrete and masonry surfaces; however, on timber decking where there is a risk of condensation or moisture forming below the sheet lead a non-woven needle-punched polyester geotextile material should be used which will not rot and allows the air to circulate. Building paper to BS 1521 class A can also be used on smooth and even surfaces as an alternative.

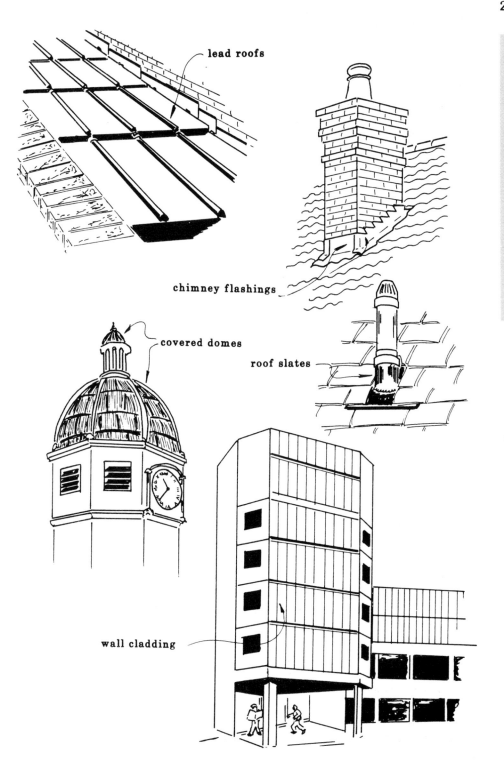

lead roofs

chimney flashings

covered domes

roof slates

wall cladding

Lead Sheet

Lead Bossing

Bossing denotes the general shaping of a malleable metal, and it applies in particular to lead. Lead can be worked into many complicated shapes although a little time, patience and practice is needed to master the skill; no book could possibly demonstrate this. A few guidelines upon the technique can, however, be given.

There are two golden rules: *rule one*: remember you are trying to move lead from one place to another, either to gain lead or lose it; you are not trying to squeeze it into one area or stretch it out to fill another. So how do we move it? The answer gives us *rule two*: set up nice curves, like waves which can flow.

Losing lead (i.e. internal corner)

First set out your work and cut off the surplus lead (see figure). The creasing lines should be set in with a setting-in stick and the sides pulled up to the required angle. About 50 mm from the edge of the corner two temporary stiffening creases are put into the sheet; this helps the corner keep its shape while bossing.

After ensuring that your work has no high ridges (if so they should be rounded), a few blows are directed with a bossing mallet in a downward direction to set in the base of the corner. With the mallet being held on the inside, supporting the work and aiming blows from the bossing stick, the corner is then worked up. The blows should be directed at the metal, drifting it outwards progressively towards the top edge of the corner (watch those high ridges; keep the curves smooth). Finish by trimming off the surplus lead which is pulled out to form a tongue.

Gaining lead (i.e. external corner)

The lead should first be folded along its length to the angle required. This will cause a hump to occur which, if it forms a ridge, must be curved, as indicated above. Then the metal is bossed around, in and down, aiming all blows towards the new corner being formed; take care not to stretch the lead.

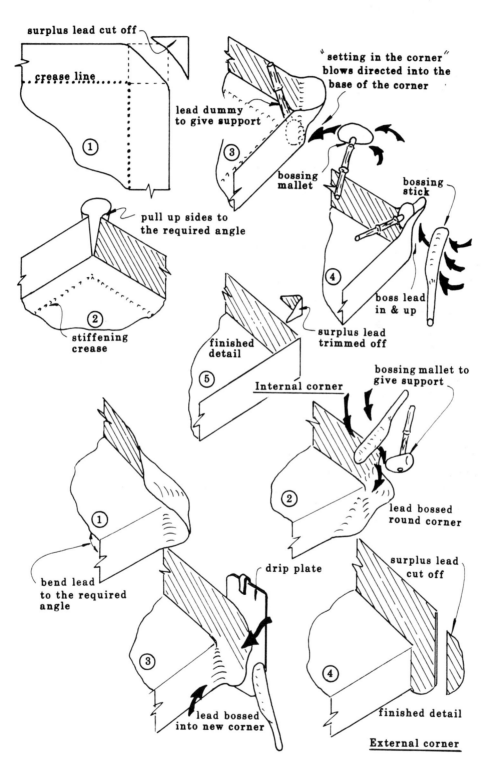

surplus lead cut off

crease line

① surplus lead cut off

"setting in the corner" blows directed into the base of the corner

lead dummy to give support

③

bossing mallet

bossing stick

④

boss lead in & up

pull up sides to the required angle

②

stiffening crease

⑤ finished detail

surplus lead trimmed off

Internal corner

bossing mallet to give support

② lead bossed round corner

drip plate

① bend lead to the required angle

③ lead bossed into new corner

surplus lead cut off

④ finished detail

External corner

Lead Bossing

Sheet Fixing

When covering any roof or wall structure with metallic and non-metallic materials it is essential to secure the sheet weathering material to the building. This is done by incorporating various fixing clips with the various details, as indicated below. In general the spacing of any fixing should be between 300 and 500 mm, depending upon exposure. It is always worth remembering that one clip too many is better than one clip too few.

Wedges A fixing used to secure an abutment flashing turned into brick or blockwork. The wedge consists of a strip of scrap lead, 20–25 mm wide, folded over several times to produce a thick section. The wedge is located in the way shown in the figure and the final front edge pointed up with cement mortar.

Clips A fixing used to secure the free edge of any lead flashing, thus preventing its lifting in high winds. Several designs of clip will be used to suit different situations, ranging from individual clips 50 mm wide to continuous fixing strips.

To give adequate strength to any clip, especially in exposed locations, it is advisable to devise the clipping system to prevent undue lifting of the material. Copper and stainless steel make the most suitable clip, giving a stronger hold; the difference in colour often gives objection; however, this can easily be remedied by tinning its exposed surface, i.e. applying a thin coating of solder. There is no problem with electrolytic corrosion in using these 'mixed' metals as they are compatible.

Secret tack A special clip which allows the middle of a large section of lead to be held in position, preventing its sagging. First a strip of lead about 100 mm wide is welded onto the back of the sheet lead, leaving a long tail or free end. The sheet lead is offered up to the building surface and the free end of the 100 mm strip is passed through a slot in the structure. Finally the free end is secured to the internal surface.

Lead dot A sheet fixing used to secure sheet lead to masonry. To achieve a lead dot fixing, first a dovetail-shaped hole is cut in the brickwork; then, once the lead has been laid, a dot mould is placed over the hole and molten lead is poured into it, filling the dovetail cavity and leaving a dome-shaped head. There are two other types of dots, these being the soldered dot and the lead welded dot; they are often used to weather the fixing, securing lead to vertical timber cheeks or cladding, and they are formed in the way shown in the figure.

Nailing Where nails are used to secure the lead at its top edge or along one side, large-headed copper clout nails should be used, a minimum of 19 mm long and having serrated shanks; these or nails of a similar design are recommended because they cannot easily be pulled out. Note that where nailing is used the fixing must be suitably covered and in all cases must not restrict the free movement of the lead.

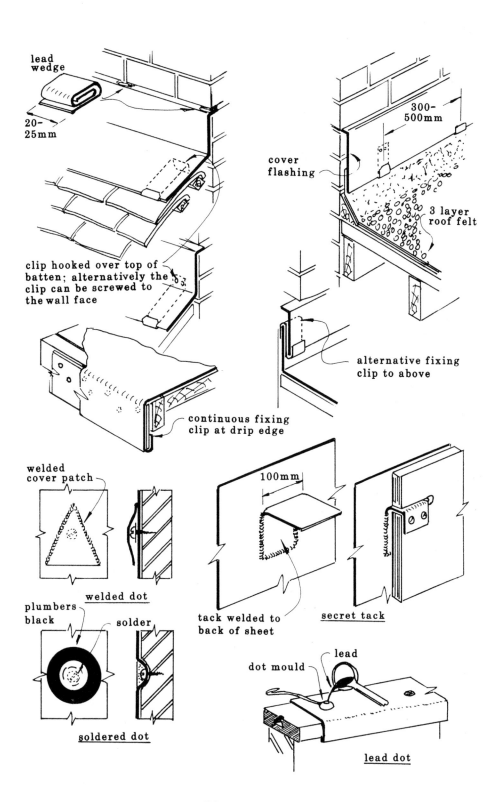

lead
wedge

20-
25mm

cover
flashing

300-
500mm

3 layer
roof felt

clip hooked over top of
batten; alternatively the
clip can be screwed to
the wall face

alternative fixing
clip to above

continuous fixing
clip at drip edge

welded
cover patch

welded dot

100mm

tack welded to
back of sheet

secret tack

plumbers
black

solder

soldered dot

lead

dot mould

lead dot

Sheet Fixing

Lead Roof Coverings and Wall Cladding

Relevant British Standard
BS 6915

The maximum sizes of individual pieces of sheet lead must be restricted to allow for the continued expansion and contraction of the material, incorporating various joints (see figure). Large roofs are divided into a series of smaller areas called bays and the size of a bay will depend upon the thickness of the sheet lead and its location (see table). However, the width of bay may be increased providing the length is reduced, in each case ensuring the maximum size area remains about the same as that indicated.

Maximum size of bay for all roof inclines

BS code No	Flat and pitched roofs up to 80°		80° pitch to vertical cladding	
	Maximum spacing for joints with fall (e.g. rolls) (mm)	Maximum length between joints (e.g. drips) (mm)	Maximum spacing for joint with fall (e.g. welts) (mm)	Maximum length between joints (e.g. laps) (mm)
4	500	1500	500	1500
5	600	2000	600	2000
6	675	2250	600	2000
7	675	2500 <10°*	650	2250
8	750	3000 <10°†	700	2250

*This distance is reduced to 2400 mm where the roof pitch is 10–60° and down to 2250 mm where the pitch is 60–80°
†This distance is reduced to 2500 mm where the roof pitch is 10–60° and down to 2250 mm where the pitch is 60–80°

Method of fixing Apart from the sheet fixings previously mentioned, the bay is generally only nailed down at the highest end at 50 mm intervals, using copper clout nails, and one-third the length of the bay nailing only to the undercloak (see figure). The rest of the lead is left to lie freely. Note that for pitched roofs and vertical cladding two rows of nails will be required (three rows for codes 7 and 8).

In addition, vertical clad surfaces sometimes require intermediate fixings such as a lead dot or secret tack; caution must be given to ensure the thermal movement is not impeded, as experience has shown that with numerous intermediate fixings fatigue cracking can result.

Preformed cladding Sometimes to assist on-site installation time, preformed lead-faced panels are produced in a workshop (see figure). The panels are simply hooked onto galvanised iron or stainless steel bars. The design of the fixing bracket is such that as one panel is engaged onto the bar its angled design pulls the panel in close to the building. Note that the joint between the sides of each panel is weathered with a lead junction with a return welt incorporated and a lap is included at the top. The panels cannot be lifted off as this is restricted by the sheet above.

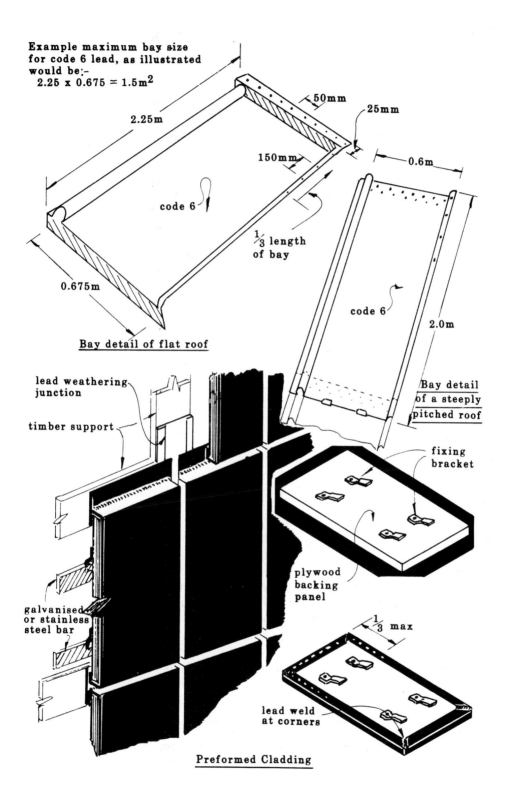

Example maximum bay size
for code 6 lead, as illustrated
would be:-
2.25 x 0.675 = 1.5m²

2.25m

50mm

25mm

150mm

0.6m

code 6

1/3 length
of bay

0.675m

code 6

2.0m

Bay detail of flat roof

Bay detail
of a steeply
pitched roof

lead weathering
junction

timber support

fixing
bracket

galvanised
or stainless
steel bar

plywood
backing
panel

1/3 max

lead weld
at corners

Preformed Cladding

Lead Roof Coverings and Wall Cladding

Expansion Joints for Lead Roofs

Relevant British Standard
BS 6915

Joints running with the fall of the roof

Wood cored roll A joint used on flat and pitched roofs using a timber roll cut to the dimensions indicated in the figure. The undercut at the base is designed to prevent the lead being lifted by high winds. The splash lap is incorporated to stiffen the free edge and keep it in position; however, where for aesthetic reasons the splash lap is undesirable it may be omitted, as long as additional copper clips, at 450 mm centres, are provided to which the overcloak has been welted.

Hollow roll An alternative roll detail which can be used, omitting the timber core. The joint can be used for all roof inclines; however, it will be subject to damage on shallow pitched roofs due to foot traffic. The hollow roll will require the inclusion of copper or stainless steel clips and is formed by turning the roll over a wood core or spring which is removed upon completion.

Welts These joints are only suitable for steeply pitched roofs where they will not be vulnerable to damage by foot traffic. As with wood cored and hollow rolls, suitable clips will need to be incorporated. Welts are also occasionally employed for joints running across the fall.

Joints running across the fall of the roof

Drip A joint used for roofs up to 10° and used in conjunction with a wood-cored roll. The drip should be 50–60 mm in height and be designed to include a splash lap which prevents the lead lifting in strong winds. The larger drip will be employed where a roll abuts the drip.

Two designs of drip are recommended: those for flat roofs up to and including 3°, in which the undercloak is rebated into the roof deck above and nailed at 50 mm spacing; and those for roofs which exceed 3°, for which the undercloak should terminate with two rows of copper clout nails at 75 mm intervals (see figure).

This second method is designed to prevent the lead slipping because of creep. Where this second method is adopted it will be necessary to seal over the nail fixings with a welded dot. In the past a 40 mm drip incorporating an anti-capillary groove was sometimes used; however, due to the fact that the groove gradually becomes blocked up with dust and dirt, etc., it proves ineffective and is also vulnerable to water penetrations in storm conditions.

Lap A joint used where the roof pitch exceeds 10°, up to and including vertical surfaces. Any lap joint will require an effective 75 mm vertical cover. The top end of the undercloak is usually fixed down with two rows of copper clout nails and the overcloak lapped to incorporate a continuous fixing clip to prevent wind lift. For roofs below 30° pitch the continuous clip usually takes the form of a piece of lead welded to the underlap. For roofs over 30° a copper or stainless fixing clip will suffice. When lap joints are employed the joints are usually staggered to avoid excessive thickness and over-complicated intersections with other joints.

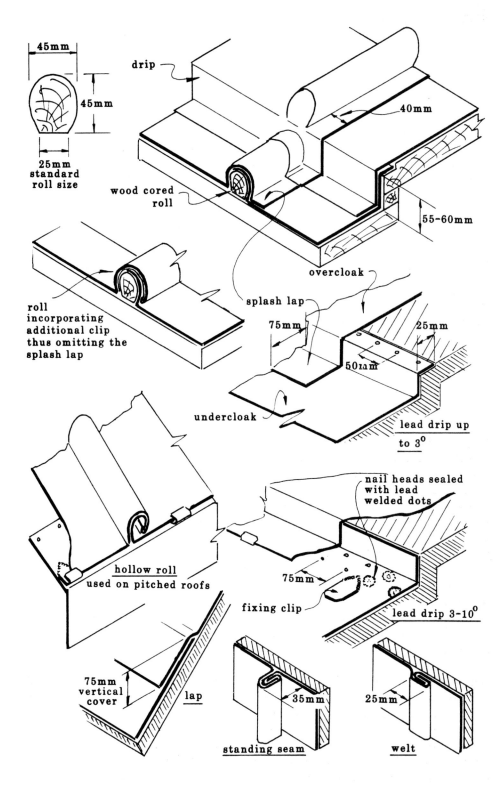

45mm

45mm

25mm
standard
roll size

drip

40mm

55–60mm

wood cored
roll

roll
incorporating
additional clip
thus omitting the
splash lap

overcloak

splash lap

75mm

25mm

50mm

undercloak

lead drip up
to 3°

nail heads sealed
with lead
welded dots

75mm

hollow roll
used on pitched roofs

fixing clip

lead drip 3–10°

75mm
vertical
cover

lap

35mm

standing seam

25mm

welt

Expansion Joints for Lead Roofs

Abutment Flashings in Lead

Cover flashing A strip of lead which is designed to weather the upstand of a roofing material such as that from a flat lead roof or bituminous felt type covering. The cover flashing is turned into the brickwork at the brick course above the upstand and allowed to hang down to give a minimum of 75 mm vertical cover. It is essential that the cover flashing is suitably clipped at 450 mm centres via lead wedges into the brick joints and hanging clips along its lower edge. The maximum length of any flashing is not to exceed 1.5 m; therefore, where lengths longer than this are needed a lap joint will be required at adjoining sections, giving a minimum of 100 mm lap (a minimum of 150 mm in exposed locations).

Apron flashing Where a pitched roof abuts a wall, a lead flashing will be required to give a minimum of 75 mm vertical upstand (see figure). The amount of cover given to lie down over the roof slope is to be a minimum of 150 mm; however, where the roof pitch is below 25° or in an exposed location this distance should be increased to 200 mm. Sufficient hanging clips are essential to prevent the lead sagging and eventually pulling the lead from the brickwork. It is usually possible to boss the lead *in situ* over the contours of most designs of roof tile; however, where sharp raised edges are encountered the lead can be cut and welded, thus avoiding excessive thinning.

Stepped flashing A special cover flashing which weathers a pitched roof to brickwork. The flashing could be either single steps or a continuous running strip, not exceeding 1.5 m in length; longer lengths are achieved by overlapping further strips. There are two basic ways in which step flashings are designed: by using the step flashing in conjunction with soakers; or by using what is called step and cover flashing.

The setting out of step flashing is best described in the figure using a piece of material 150 mm wide, which allows for a water line of at least 65 mm. In the case of step and cover flashing add to this width 150–200 mm, depending upon the tile profile and its exposure, to lay out across the tiles.

Soakers The soaker is designed to give an upstand at the abutment of a pitched roof to a wall, being incorporated with the laying of the tiles or slates, and is made on site from sheet lead.

The width of a soaker used to join up to an abutment must be a minimum of 175 mm; this would allow for an upstand of 75 mm against the wall and the remaining 100 mm can lay out across the roof and under the tiles or slates. The length of the soaker varies and depends upon the length of the roof slates. The calculation used to find the length is: gauge + lap + 25 mm. The gauge is the distance between the roof slate battens; the lap is the distance one slate overlaps the slate next but one below it; and the 25 mm is optional and is purely for fixing purposes, being bent over the battens to prevent their slipping.

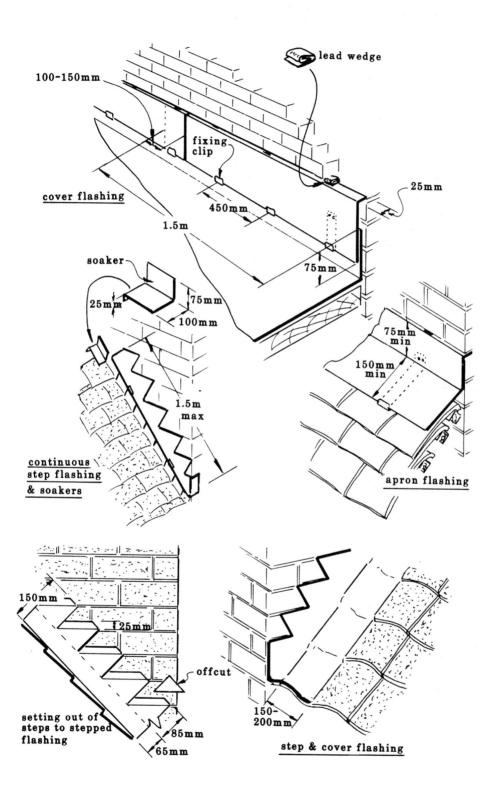

lead wedge

100-150mm

fixing clip

cover flashing

25mm

450mm

1.5m

75mm

soaker

25mm

75mm

100mm

1.5m max

75mm min

150mm min

continuous step flashing & soakers

apron flashing

150mm

25mm

offcut

setting out of steps to stepped flashing

85mm

65mm

150-200mm

step & cover flashing

Abutment Flashings in Lead

Chimney Flashings in Lead

The roof weathering which prevents rain-water penetrating the building where a chimney stack passes through the roof structure. Chimney flashings consist of: a front apron; soakers; step flashing; back gutter; and a cover flashing. With all sheet roof work the lowest pieces are positioned first and one works up the roof allowing the higher pieces to lap over those below.

Front apron The lowest piece, being made from a piece of lead 300–350 mm wide; its length is the width of the stack plus 300 mm, allowing 150 mm each side of the roof. In some situations 200 mm may be required on each side to allow for deep-profiled tiles or exposure. The front apron can be bossed or welded (see figure) to produce a detail which maintains the required fixing and weathering cover, as previously described.

Stepped flashing and soakers In addition to what has been said already (see page 306) the front edge is turned round the chimney-stack to give a 75 mm cover along the corner detail. It is possible to finish the side flashing at the corner, i.e. not turning the front edge, provided the previous front apron maintains a 150 mm minimum turn to the stack.

Back gutter As with the front apron it is possible to boss this detail; however, it does involve considerably more work than that for the apron. The approximate width of material is 500 mm, allowing for 100 mm minimum upstand to the brickwork, 150 mm for the gutter sole and the remaining to lay up under the tiles or slates. The length is the same as for the front apron. As with the front apron the details are best identified from the figure.

The final, and highest, piece of chimney flashing is the cover flashing which is designed to cover the vertical upstand of the back gutter. Cover flashings are described on page 306.

Sometimes the chimney is located at the apex of the roof (ridge), in which case no back gutter will be required; instead the stepped flashing is terminated at the ridge and a saddle piece used to provide suitable cover at the highest point.

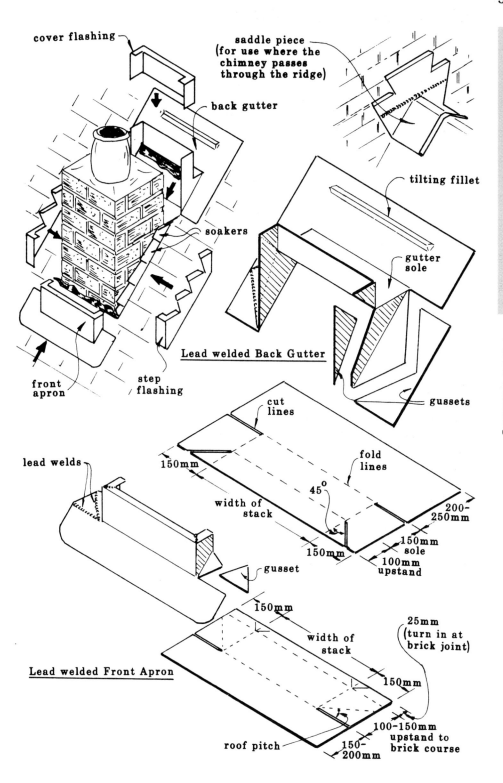

9 Sheet Weathering

cover flashing

saddle piece
(for use where the
chimney passes
through the ridge)

back gutter

tilting fillet

soakers

gutter
sole

Lead welded Back Gutter

gussets

front
apron

step
flashing

cut
lines

fold
lines

lead welds

45°

200-
250mm

150mm

width of
stack

150mm
sole
100mm
upstand

150mm

gusset

150mm

25mm
(turn in at
brick joint)

width of
stack

150mm

Lead welded Front Apron

100-150mm
upstand to
brick course

roof pitch

150-
200mm

Chimney Flashings in Lead

Lead Slates and Pitched Valley Gutters

Lead slates A special flashing used to weather any obstruction passing through a roof, generally found to be a ventilating or flue pipe. The size of the lead required to produce the slate will be dependent upon the roof pitch, the size of the pipe and the type of roof tiles. Generally the slate should be made to give 150 mm minimum distance cover at all points, i.e. sides, back and up the pipe itself (see figure).

When positioning the slate the gap between the pipe and the lead also needs to be weathered; this is achieved by using a solvent welded collar for plastic pipe or by dressing the lead into the top of the pipe opening. Sometimes a mastic or silicone sealant is used to make this seal, although only as a last resort.

The lead slate can be formed by bossing although it would be much more practicable to lead weld as necessary. Firstly a piece of lead wide enough is turned around a piece of rigid pipe and its meeting edges are butt welded together. The pipe is then cut to the required pitch, positioning the weld to the back; the raking edge is dressed to form a flange which is placed on the base. The hole is marked and cut and the sleeve is lap-welded as necessary.

Sometimes overflow pipes passing through a steeply pitched roof need to be weathered in a similar fashion to the slate. This is known as a lead sleeve.

Pitched valley gutters Where a pitched roof requires to turn an internal angle the intersection will need to be weathered. Several methods can be adopted for this purpose, one such design using lead linings laid onto valley boards. For most gutters the valley boards should extend 225 mm up each side from the centre of the gutter, with a tilting fillet positioned 150 mm up from the base.

The tilting fillet is designed to tilt the free edge of the tile or slate less steeply than the rest of the roof to ensure that the tiles or slates bed tightly on one another; it also assists in the prevention of capillary attraction between the sheet metal and the general roof covering material. A gap of 125 mm should be maintained between the two roof pitches to prevent the build-up of leaves, etc., which may cause a blockage.

Where bituminous sarking felt is used beneath the tiles it should not be allowed into the sole of the gutter, below the lead because, should the felt soften in hot weather the lead may stick to the substrate, restricting its free movement. The fixing of the lead itself should be limited to two rows of copper nails across the top of each piece and on gutters over 60° additional fixing used only at the sides of the top third. No piece of lead should exceed 1.5 m in length and a 150 mm minimum lap is required at each piece, this distance being increased to 220 mm where the roof pitch is lower than 20°. Should two valleys meet, a saddle will be required at the highest point (see figure).

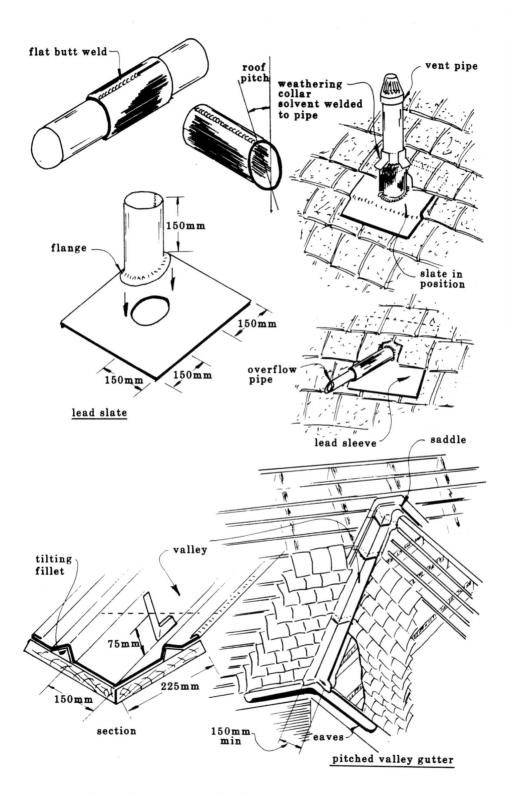

flat butt weld

roof pitch

weathering collar solvent welded to pipe

vent pipe

slate in position

flange

150mm

150mm

150mm

150mm

lead slate

overflow pipe

lead sleeve

saddle

tilting fillet

valley

75mm

225mm

150mm

section

150mm min

eaves

pitched valley gutter

Lead Slates and Pitched Valley Gutters

Gutter Linings

Where a lead covered roof terminates at its lowest end a sheet lead gutter will usually be formed. Lead-lined gutters will also be found where a tiled or slated roof is designed with a parapet wall or where two pitched roofs meet at a horizontal valley. The fabrication of the gutter is for the most part identical to the installation of a small flat roof, the size of bay being that previously identified for lead roof coverings.

Drip design The splash lap of a drip may be omitted in gutter linings because the detail will be less vulnerable to lifting in strong winds. It is often desirable to omit this splash lap, not only in order to simplify installation but also because grit and dirt tend to collect under the lap, which slows the flow of water through the gutter. Where the splash lap is omitted it is essential that the side lap is maintained.

Gutter design The position of drips in a gutter will be dependent on the sheet thickness which is often governed by the outlets and amount of fall. For new work the designer will allow sufficient height for the required fall to suit the drip spacing for a given thickness of lead. However, with existing gutters the outlets are usually fixed and the amount of height for the fall governed by the existing wall height. Thus it may be necessary to choose a thicker lead size, allowing for longer bay lengths (see figure).

Box gutters A gutter lining, usually with parallel sides, with a minimum width of 225 mm. All upstands from the gutter must be a minimum height of 100 mm. Where it meets a tiled or slated roof the lead is carried under the tiles up the roof slope to maintain the minimum 100 mm depth. It is permissible to increase the upstand to 200 mm for codes 5 and 6 or 300 mm for codes 7 and 8 where necessary, but with excessively deep-sided box gutters a separate piece of lead will be required. It is important not to nail the taller upstands as it will restrict their free thermal movement.

Tapered gutters A tapered gutter will occur where two pitched roofs meet or where a pitched roof abuts a vertical wall. A tapered gutter is distinguished by the sole (base) of the gutter becoming progressively wider at each drip, becoming its widest at the highest point. The minimum width at the lowest end should be 150 mm and for long gutters a roll, or several rolls, may be required to divide excessively wide sections into two or more bays.

Outlets

Chute A chute outlet is where the sole of the gutter is taken through the parapet wall to discharge into a hopper head. This form of outlet is not prone to blockages by leaves, etc., and allows for a good flow capacity.

Catchpit A design of outlet which allows for an internal downpipe. The minimum depth should be 150 mm and its length equal to that of the width of the gutter. Smaller depths can be used provided the length is increased to prevent surcharging in storm conditions. When a catchpit is chosen it should incorporate an overflow to warn of blockages.

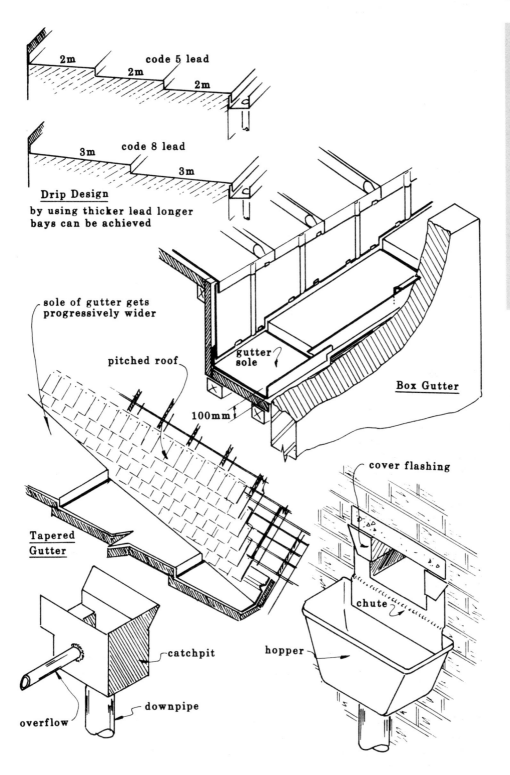

2m   code 5 lead
2m
2m

code 8 lead
3m
3m

Drip Design
by using thicker lead longer
bays can be achieved

sole of gutter gets
progressively wider

gutter
sole

Box Gutter

pitched roof

100mm

cover flashing

**Tapered
Gutter**

chute

catchpit

hopper

overflow

downpipe

Gutter Linings

Dormer Windows

The dormer window is a typical leadwork detail which will be found as a common architectural feature in a pitched roof. It involves developing the skills previously described and adapting processes to form a covering consisting of the following three parts.

Tray and apron flashing A detail in which the lead is bossed or welded to fit at the base of the window opening. The window sill sits onto the tray. The lead extends out to lie down the roof slope and around the front face and side of the vertical timber (see figure). The apron detail is similar to that of the chimney front apron.

Side cheek This is the side cladding on which the lead hangs vertically. The most important consideration to observe is the size of the panel and its fixing, which the page on roof cladding should be reviewed (see page 302). The lead is turned round the front edge and a return welt formed upon which a timber covering bead is placed.

The base of the side cheek is weathered either using soakers, as necessary where the dormer protrudes from a tiled or slated roof, or weathered using a continuous side flashing. Where a side flashing is used it is either welted to a lead covering, should this be the main roof covering material, or it is simply laid out across the roof a minimum distance of 150 mm, and fixed as necessary. **Note**: the maximum flashing lengths are not to exceed 1.5 m.

Dormer top The top of the dormer is covered in the same way as any small flat roof, being divided with wood-cored rolls if necessary. Where the lead terminates at the sides and front of the top it should be turned down to give a minimum 75 mm vertical cover to the side cheek, fixing clips being incorporated at 450 mm centres.

The back lays up under the roof covering above (usually 200 mm, minimum) depending upon the roof pitch. Generally the direction of fall from the dormer top is made to incline to the sides, although a small dormer may have no fall at all. However, where the dormer is large, comprising three or more bays, the fall will need to flow to the front where an eaves gutter may be positioned to collect the water run-off – or directed backwards, in which case a gutter lining will be needed.

As an alternative to a flat-topped dormer, a curved top may be designed for small or medium sized roofs. This may overcome the need for a front or rear gutter. Where a curved top is included the sheets of lead are generally joined using welts running with the curve over the dormer top. Should the length of the curve exceed 1.5 m a cross joint will be needed.

Sometimes an internal or inset dormer will be found. These may be dealt with in ways which amount to a simple variation of those described above.

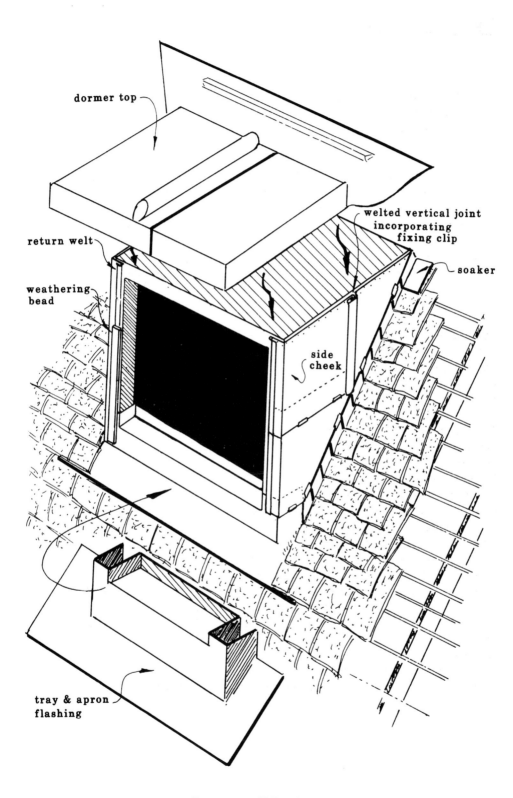

dormer top

welted vertical joint
incorporating
fixing clip

soaker

return welt

weathering
bead

side
cheek

tray & apron
flashing

Dormer Windows

Aluminium and Copper Sheet Weathering

Relevant British Standard
CP 143

Aluminium sheet This can be obtained in rolls with various widths, ranging from 150 to 900 mm. There are various thicknesses of the material available, these ranging from 0.6 mm to 1 mm. Only two grades of aluminium are available for roofwork; these are: super purity, which contains 99.9% aluminium; and commercial purity, which contains 99% aluminium, the other 1% being made up of other elements. Aluminium has similar working properties to those of copper, and in general roof details would be carried out along the same lines as those for copper.

Copper sheet This can be obtained in almost any size and in a range of thicknesses, although the standard sizes usually specified are 1830 mm x 610 mm, which should be fully annealed for general roofwork. Sheet copper remains unaffected by changes in temperature; and it will not creep but become work-hardened if cold worked too much. The method of jointing sheets of copper together varies depending on the circumstances and specifications.

Method of jointing In general for joints which run across the flow of water (transverse joints) the method of jointing is either by drips, double lock welts or single lock welts (see table). Where a roof pitch is between 6 and 20° if a double lock welt is used it must be sealed, either with boiled linseed oil or a non-setting mastic.

Recommended transverse joints to be used

Jointing method	Roof pitch
Drips	up to 6°
Double lock welt (sealed)	6–20°
Double lock welt	20–45°
Single lock welt	over 45°

For joints which run in the same direction as the flow of water, standing seams or batten rolls are used. In the case of roofs on which foot traffic is expected standing seams should not be used. This should prevent any flattening of the joint if it is trodden on.

The dimensions for single lock and double lock welts are open to debate and in most cases plumbers make them of a size to suit their forming blocks, or the special grooving tools which assist them in making the joint. Sometimes the size of the welt is specified by the client, in which case the client's requirements must be observed.

The sizes given in the figure are sizes I have found suitable in the past and are used on materials such as copper and aluminium. Holding-down clips should be incorporated with any welted seams and spaced at 300–500 mm centres, thus giving a secure detail.

When completing any details in sheet aluminium or copper it is essential to remember that, as with lead, the area of the roof must be minimised to allow for the thermal movement of the sheet; and large roofs will need to be divided into bays. In many cases the previous notes for sheet lead can be adapted to suit these roofing materials, e.g. replacing the weld with a welted seam.

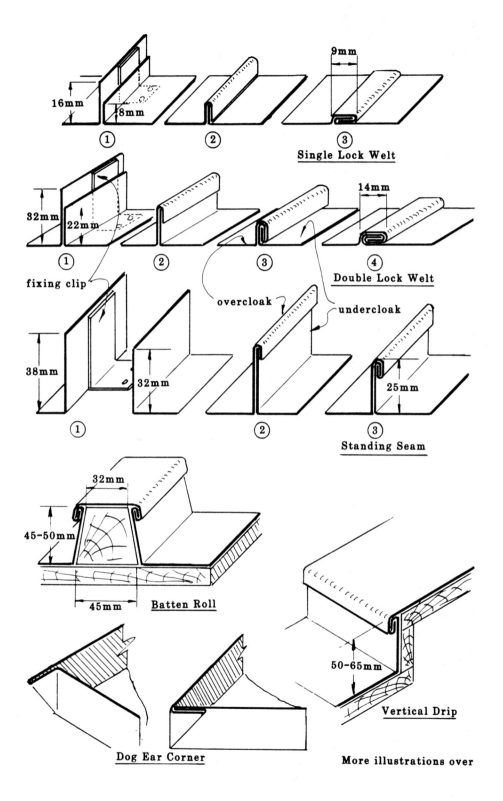

16mm
8mm

① ② ③
Single Lock Welt

9mm

32mm
22mm

① ② ③ ④
Double Lock Welt

fixing clip

overcloak
undercloak

14mm

38mm
32mm
25mm

① ② ③
Standing Seam

32mm
45–50mm
45mm **Batten Roll**

50–65mm
Vertical Drip

Dog Ear Corner

More illustrations over

Aluminium and Copper Sheet Weathering 1

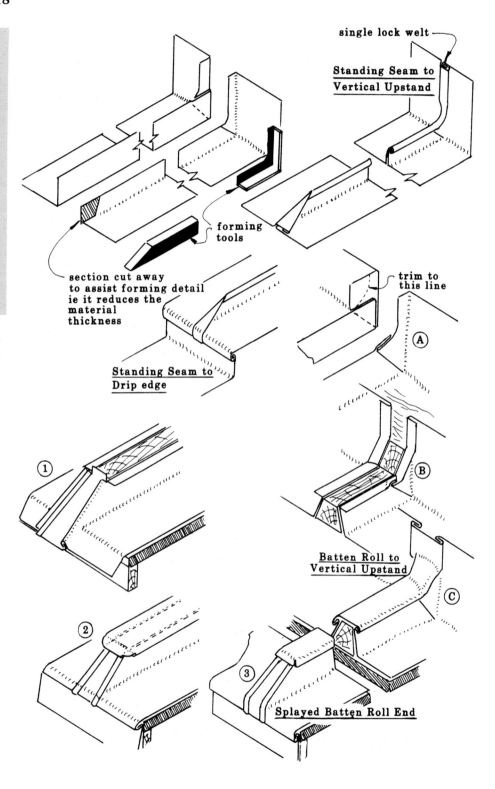

single lock welt

**Standing Seam to
Vertical Upstand**

forming
tools

section cut away
to assist forming detail
ie it reduces the
material
thickness

trim to
this line

**Standing Seam to
Drip edge**

Ⓐ

①

Ⓑ

**Batten Roll to
Vertical Upstand**

②

Ⓒ

③

Splayed Batten Roll End

Aluminium and Copper Sheet Weathering 2

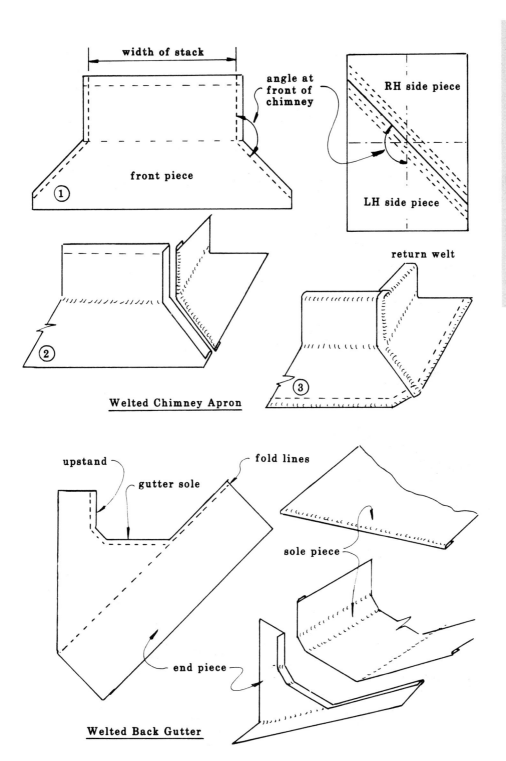

width of stack

angle at front of chimney

RH side piece

LH side piece

front piece

①

return welt

②

③

Welted Chimney Apron

9 Sheet Weathering

upstand

gutter sole

fold lines

sole piece

end piece

Welted Back Gutter

Aluminium and Copper Sheet Weathering 3

Stainless Steel and Zinc Sheet

Relevant British Standards
CP 143 and BS 6561

Stainless steel has been widely employed for roofing on the continent, particularly in France, since the early 1950s, but only since 1974 has the material been used in this country. One of the main advantages of stainless steel as a roofing material is its very low scrap value; thus it does not present an attractive proposition to the thief and is likely to remain on the roof for 100 years or more.

The methods employed to lay stainless steel to a structure are the same as those for the zinc roll cap system. This material is obtained either with a low reflective finish (a silvery grey colour) or it can be obtained with a 'solder coating' finish (called a terne coated finish) designed to weather to a dull lead-like appearance. Stainless steel and lead are chemically compatible, so a stainless steel roof can be dressed with lead flashing without fear of corrosion to either metal.

Zinc sheet of commercial quality was available for many years in the UK but during the 1960s zinc alloys were developed, and these have since replaced the ordinary commercial zinc.

Two alloys are now available. Type A, a zinc, titanium and copper alloy, is used for general roof coverings; it is supplied in sheets 1 m wide and up to 3 m long. Type B is a zinc/lead alloy primarily used for flashings or small roofing applications, e.g. canopies. Type B is supplied in coils of several widths. The usual thicknesses of the sheet for roofwork are 0.7 and 0.8 mm.

The working properties of zinc are not as good as those of such materials as aluminium, copper and lead; therefore a special design of roofwork is employed called the 'roll cap system', although it is possible to join the sheets using the standing seam method. Where zinc alloys are used many details comparable to those possible with copper and aluminium may be achieved, although generally larger welts are required.

Roll cap system

This is a specially designed method of laying the sheet which allows for the jointing of the material without the need to form tight fitting welts. For joints which run across the flow of water (transverse joints) the method of jointing is either by drips, or for roofs with a pitch greater than 10°, mastic sealed single lock welts may be used with a welt size of 45 mm. Where the pitch is over 25° the overcloak can be reduced to 30 mm, omitting the sealing compound. The joints which run with the fall of the roof are made with a special design of batten roll (see figure), the capping piece being held down by a specially formed clip. The drips may be formed in either of two ways, as a welt or a beaded edge. The welted edge is usually the easiest to form. Note that the welt is not turned down as with copper drips.

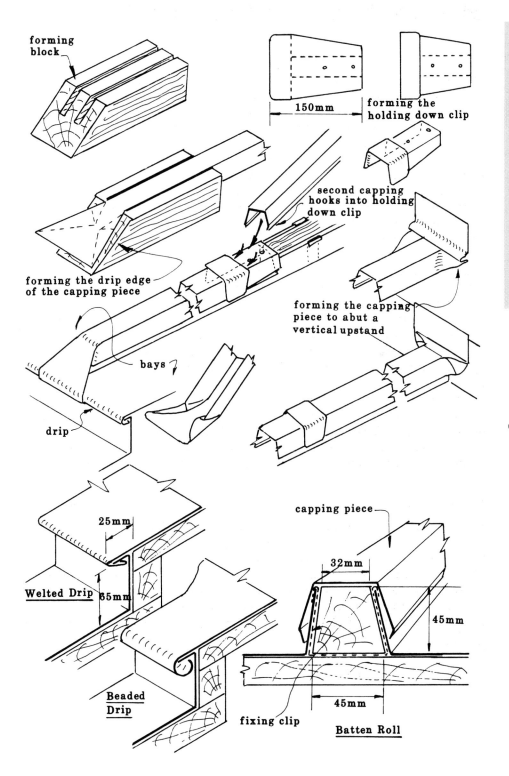

forming
block

150mm

forming the
holding down clip

second capping
hooks into holding
down clip

forming the drip edge
of the capping piece

forming the capping
piece to abut a
vertical upstand

bays

drip

25mm

Welted Drip 65mm

Beaded
Drip

capping piece

32mm

45mm

fixing clip

45mm

Batten Roll

Stainless Steel and Zinc Sheet

Non-metallic Sheet Weathering

For many years the plumber has been occasionally called upon to lay a roof weathering using a material called Nuralite – a trade name for an asphalt bonded asbestos roofing sheet of laminar construction. Nuralite sheet is available in various forms, including a grade which does not contain any asbestos at all and is ideal in areas where asbestos is prohibited. It is referred to as Nuralite Nutec. There is a further grade known as Nuralite FX sheet which is much easier to fabricate and delaminate. The FX sheet has other advantages over the standard nuralite sheet in that it will withstand more readily the movement experienced in new building work, and tends to have a better resistance to general wear and tear. The standard sheet size is 2400 mm × 900 mm. It weighs about 2.45 kg/m^2.

Nuralite is easy to cut with either tin snips or a knife, and is usually worked by heat from a blowlamp; this softens the material for a period to allow one to mould it to the required shape. Once the plumber has it in its softened state he needs to master only two basic details: the internal and external corners (the internal corner being constructed forming a dog ear, see figure).

The jointing of Nuralite is very simply made by one of three basic methods: the D12C joint, the lap joint or the delaminated joint. The third jointing method, the delaminated joint, is restricted to external corners. The lap joint is normally restricted to sloping roofs and when making this joint a 75 mm lap would be required with the application of a No 3 jointing compound (see table).

When laying Nuralite roofing sheet to roofs the sheets are generally just laid onto the decking with a 2 mm gap between each sheet. One side of the Nuralite is smooth compared to the other and it is this smooth side that should be uppermost and exposed to the atmosphere owing to its superior water-shedding properties. The Nuralite should be fixed to the decking with either a No 10 or No 30 jointing compound. The edge of the sheet should be nailed at 300 mm centres, 12 mm from the edge of the sheet.

Once the sheets have been laid and all the edges and sides completed the joint is simply made by pre-heating up the pre-coated D12C jointing strip with a blowlamp until the bitumen is moist. The strip is laid onto the butted edges and consolidated at the edges with a curved block, or a specially designed roller. Finally any excess bitumen is removed with a scraper. When covering a structure with this method the joints which run across the flow should be completed first.

Nuralite can be laid in the more traditional methods of roll-cap roofing as would be employed on zinc roofs but this method is only usually carried out for reasons of appearance. There is a range of pre-formed flashings which is available to meet many common applications and more details can be obtained from the Nuralite company in Rochester, Kent.

Nuralite jointing compounds

Compound no.	Description
1	A bitumastic welding block used for delaminated joints
3	A bitumastic block compound used for lap joints
10	A cold bitumastic undersheet adhesive used on roofs up to 40° pitch
30	A hot bitumastic undersheet adhesive used on roofs under 15° pitch or where high wind pressures are likely
D12C	A 100 mm wide heat-applied jointing strip

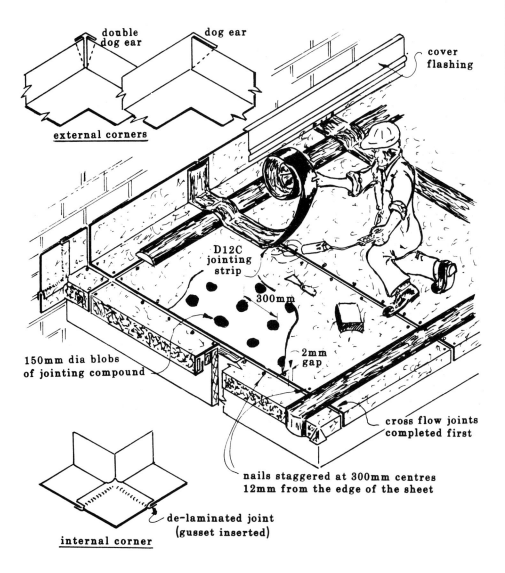

double dog ear

dog ear

cover flashing

external corners

D12C jointing strip

300mm

150mm dia blobs of jointing compound

2mm gap

cross flow joints completed first

nails staggered at 300mm centres 12mm from the edge of the sheet

de-laminated joint (gusset inserted)

internal corner

Non-metallic Sheet Weathering

Part 10

Assessing Your Knowledge

This section of the book is designed to allow you to check your level of knowledge of the skills of plumbing. The assessments have been divided into three areas as follows:

(1) **Self assessment:** 100 multiple choice questions on pages 326–40. Answers are to be found at the end of the book (page 391).
(2) **Supplementary assessment:** 100 short-answer questions (pages 341–80) typical of the sort you are likely to be given in college-based training programmes leading to an NVQ in plumbing. No direct answer is given here to any question, although the answer to each lies somewhere in the book. To a large extent the questions follow the order in which information is to be found in this book; hence it is possible to work through the book to find the answers. However, it is probably best to tackle the questions as your knowledge grows. Your answers can be confirmed correct by your college tutor who may decide you can use this as evidence in your NVQ portfolio.
(3) **Problem solving:** ten short problems (pages 381–390) of a sort typical of those encountered in plumbing systems; a trained plumber should be able to find the solutions without too much difficulty. The answers to these questions may need some thought and transference of knowledge, or indeed consultation with others, such as your college tutor.

Self Assessment

(100 multiple choice questions)

Check your responses to these questions by referring to the answer sheet on page 391.

The main type of multiple choice question will be of a type known as the 'four option multiple choice'. This comprises either a direct question or statement, known as the stem, followed by a choice of four different answers, called the responses. Only one of these responses is the correct answer (the key); the others are incorrect (but plausible) distractors to the key. Candidates are required to select their response by choosing either (a), (b), (c) or (d).

Example:

When running an overflow pipe from a storage cistern it is necessary to:

(a) Ensure a trap is fitted in the pipe-run.
(b) Ensure the connection to the drain is watertight.
(c) Discharge the pipe into a gutter, if possible.
(d) Terminate the pipe in a conspicuous position.

d

The correct answer is to terminate the pipe in a conspicuous position; therefore the response should be (d).

(1) Which of the following is a non-ferrous metal?

(a) Aluminium.
(b) Cast Iron.
(c) Low Carbon Steel.
(d) Wrought Iron.

(2) What is meant by the term plumbo-solvent?

(a) The use of lead-free solder fittings.
(b) The addition of chlorine to the water supply.
(c) The ability of water to dissolve lead.
(d) The ability of water to dissolve copper.

(3) What is meant when water is said to be at its maximum density?

(a) It will expand no more.
(b) It exerts its greatest amount of pressure.
(c) It occupies the least amount of volume.
(d) It is at its highest temperature.

(4) At what temperature is the maximum density of water?

(a) −273°C.
(b) 0°C.
(c) 4°C.
(d) 100°C.

(5) Which of the following metals is an alloy?

(a) Brass.
(b) Iron.
(c) Aluminium.
(d) Copper.

(6) Heavy grade low carbon steel tube to BS 1387 is identified by which colour coding?

(a) Red.
(b) Blue.
(c) Brown.
(d) Green.

(7) Electrolytic corrosion (galvanic action) occurs when:

(a) Two dissimilar metals are in contact with an inert gas.
(b) Two dissimilar metals are in contact via an electrolyte.
(c) A metal is exposed to stormy weather conditions.
(d) Two dissimilar metals are placed in a liquid incapable of passing an electric current.

(8) One method of joining to polyethylene pipe is carried out by:

(a) Solvent weld cement.
(b) Fusion weld cement.
(c) Compression joints with inserts.
(d) Capillary joints.

(9) 28 mm diameter copper tube, when fixed horizontally, should be supported at intervals of:

(a) 0.5 m.
(b) 1 m.
(c) 1.8 m.
(d) 2.7 m.

(10) Bends are often used on pipes rather than elbows because:

(a) The frictional resistance is reduced.
(b) The frictional resistance is increased.
(c) The pipework is easier to run.
(d) Insulation material is easier to apply.

(11) The type of thread cut onto low carbon steel tube is:

(a) British Standard Whitworth.
(b) British Standard Pipe thread.
(c) British Standard Fine.
(d) Metric.

(12) Annealing a metal will:

(a) Make it softer and more workable.
(b) Give it a protective coating of aluminium.
(c) Increase its tensile strength.
(d) Remove surface blemishes.

(13) Rippling to bends on copper tube when using a bending machine is caused by:

(a) The backguide being adjusted too tightly.
(b) The backguide being adjusted too loosely.
(c) The bend being pulled too quickly.
(d) Loss of hydraulic fluid.

(14) Copper and galvanised steel should not be used in the same water system because:

(a) Limescale will form in the pipework.
(b) A fungus will grow within the system.
(c) Electrolytic corrosion will occur.
(d) Erosion of the galvanised steel will occur.

(15) The term calorific value of a fuel relates to the:

(a) Energy produced from a given quantity of fuel.
(b) Weight of the fuel compared to that of water.
(c) Temperature at which the fuel burns.
(d) Gas air ratio.

(16) Which of the following is the most important commissioning procedure to carry out upon completion of a hot or cold water installation:

(a) Tidy up and ask the client to inspect the work.
(b) Work out the final account to be paid for the work.
(c) Fit the insulation material to ensure compliance with local water bye-laws.
(d) Check for leaks, adjust water levels and check flow rates.

(17) Continuity bonding is carried out to pipework to:

(a) Ensure that any earth leakage is conveyed to the ground via the water pipes.
(b) Prevent overloading of the household electrical system.
(c) Prevent an electric shock should a live wire touch any pipe.
(d) Reduce the accumulation of scale in water pipes.

(18) The correct procedure to adopt when lifting heavy objects is:

(a) Feet together and lift using the back to take all the strain.
(b) Feet slightly apart, bend at the knees and keep a straight back.
(c) Feet wide apart and lift using the back to take all the strain.
(d) Feet slightly apart, bend at the knees and keep a curved back.

(19) The person who is employed by the local authority to ensure that the Building Regulations are observed is called the:

(a) Ombudsman.
(b) Clerk of works.
(c) Building control officer.
(d) Local bye-laws officer.

(20) The colour code of wires to a 13 amp 3-pin plug is:

	Live	Neutral	Earth
(a)	Blue	Brown	Green
(b)	Red	Blue	Green & Yellow
(c)	Brown	Blue	Green & Yellow
(d)	Black	Red	Green

(21) Which of the following should not be used on fires involving electrical equipment?

(a) Foam.
(b) Dry powder.
(c) Carbon dioxide (CO_2).
(d) Vaporising liquid.

(22) One litre of water weighs:

(a) One kilogram (1 kg).
(b) One gramme (1 g).
(c) One pound (1 lb).
(d) More when heated.

(23) To obtain a suitable soft-soldered capillary joint on copper tube the joint should:

(a) Have a small space between the adjoining surfaces.
(b) Have a large space between the adjoining surfaces.
(c) Be thoroughly cleaned with emery paper.
(d) Be suitably belled out to allow a molten pool to be formed.

(24) A flux is used to:

(a) Burn into the base metal to ensure the complete removal of all oxides and dirt, etc.
(b) Assist the insertion of the pipe into the fitting.
(c) Prevent the solder vaporising and escaping from the joint.
(d) Prevent the oxidation of the metal to be soldered.

(25) When bronze welding it is necessary to use a:

(a) Neutral flame.
(b) Carburising flame.
(c) Oxidising flame.
(d) Large spreading flame.

10 Assessing Your Knowledge

(26) The type of thread found on acetylene welding equipment is:

(a) BSP.
(b) Recessed.
(c) Right-handed.
(d) Left-handed.

(27) Acetylene gas should never be conveyed in pipelines made of or incorporating:

(a) Copper.
(b) Mild steel.
(c) Cast iron.
(d) Rubber.

(28) A leak from a propane gas blow torch, or similar equipment using fuel gas, should be found using a:

(a) Match or lighted taper.
(b) Manometer.
(c) Soap solution.
(d) Coloured die, added to the gas.

(29) The process of bronze welding can be said to be a type of:

(a) Hard soldering.
(b) Soft soldering.
(c) Brazing.
(d) Autogenous welding.

(30) When welding low carbon steel or lead, using a neutral flame:

(a) Flux must be applied to the cleaned joint.
(b) The filler rod must contain the necessary flux.
(c) No flux is required.
(d) A porous weld would result.

(31) Hard waters contain:

(a) Calcium salts.
(b) Epsom salts.
(c) Acidic salts.
(d) Saline salts.

(32) An air gap must always be maintained when supplying a sink with water from a supply pipe in order to prevent:

(a) Back-siphonage.
(b) Self-siphonage.
(c) Airation of the water supply.
(d) Legionnaires disease.

(33) Where different pressures are at the hot and cold connections to a mixer tap serving a sink or bath, the tap should:

(a) Be fitted with a pressure-reducing valve.
(b) Have a divided outlet.
(c) Have a combined outlet.
(d) Be of a mono-block design.

(34) The service valve fitted on the cold distribution pipe from a storage cistern should be a fullway gate-valve in order to:

(a) Make installation easy.
(b) Prevent water hammer.
(c) Prevent blockages due to debris from the cistern.
(d) Keep frictional resistance to a minimum.

(35) When water freezes it:

(a) Expands about 1600 times.
(b) Decreases in size by about one-tenth.
(c) Increases in size by about one-tenth.
(d) Goes through a chemical change.

(36) An underground water supply pipe should be a minimum depth below ground of:

(a) 350 mm.
(b) 750 mm.
(c) 1000 mm.
(d) 7500 mm.

(37) Should the ballfloat of the float operated valve in a flushing cistern become full of water:

(a) The cistern would continuously flush.
(b) The cistern would fail to flush.
(c) Water would discharge from the overflow.
(d) Water hammer would be apparent.

(38) The main difference between a high pressure float operated valve and a low pressure float operated valve (both to BS 1212) is:

(a) The size of the ballfloat.
(b) The size of the valve body.
(c) The size of the seating orifice.
(d) The length of the lever arm.

(39) When running a hot water pipe through a wall it should be sleeved to:

(a) Allow for thermal movement of the pipe.
(b) Allow for thermal movement of the wall.
(c) Allow settlement of the building.
(d) Make maintenance easier.

(40) Secondary circulation is carried out to convey hot water to:

(a) The cylinder from the boiler.
(b) A bathroom towel rail.
(c) A point close to the draw-off point, thus preventing a wastage of water.
(d) Radiators.

(41) The feed cistern supplies water to:

(a) The cold distribution pipework.
(b) The hot storage vessel.
(c) A WC pan or urinal.
(d) A set of booster pumps.

(42) Primary circulation is the water which circulates:

(a) Between the cold feed cistern and the dhw cylinder.
(b) Between the hot storage vessel and the boiler.
(c) Between the hot storage vessel and the furthest draw-off point.
(d) Around inside the hot water cylinder.

(43) A dhw system in which the water drawn off from the taps has passed through the boiler is known as a:

(a) Primary system.
(b) Single feed system.
(c) Direct system.
(d) Indirect system.

(44) Unvented dhw supply refers to a system with:

(a) A hot stored capacity less than 15 litres.
(b) A hot stored capacity greater than 15 litres.
(c) A single feed cylinder.
(d) A storage cistern vented to the atmosphere.

(45) The water in the primary flow and return is prevented from mixing with the water in a single feed indirect cylinder by:

(a) The trapping of an air pocket.
(b) Ensuring the system has a good head pressure.
(c) Fitting a pump on the primary flow.
(d) Having separate cold feed connections.

(46) A secondary return should be returned to the:

(a) Boiler.
(b) Top third of the hot storage cylinder.
(c) Bottom third of the hot storage cylinder.
(d) Last radiator.

(47) In a vented low temperature hot water heating system the position of the feed and expansion cistern determines the:

(a) Static head.
(b) Circulating head.
(c) Circulating pressure.
(d) Amount of water which will expand upon heating.

(48) An air separator is a device which:

(a) Allows cold feed and vent connections to be closely grouped in a fully pumped system and permits air bubbles to form which simply raise up out of the system.
(b) Allows the cold feed and vent connections to be made anywhere in the system as any air drawn into the system is expelled.
(c) Prevents boiler noises.
(d) Is fitted to condensing boilers to allow the condensate to be expelled from the flue gases.

(49) A combination boiler:

(a) Allows the dhw and c.h. waters to mix.
(b) Heats both dhw and c.h. waters independently.
(c) Is a boiler which uses two fuels (e.g. gas and oil).
(d) Is a design of boiler used for sealed heating systems.

(50) When calculating the heat loss through the building fabric the external temperature is usually taken to be:

(a) 4°C.
(b) 0°C.
(c) −1°C.
(d) Irrelevant.

(51) The transference of heat by convection currents is particularly important in:

(a) Space heating.
(b) Soldering copper tubes.
(c) Refrigerators.
(d) 'U' value calculations.

(52) When a new heating installation has been completed it should be commissioned by the:

(a) Building control officer.
(b) Householder.
(c) Site agent.
(d) Installer.

(53) The size and therefore capacity of a feed and expansion cistern is based upon the:

(a) Amount of water in the heating system.
(b) Amount of water in the hot storage vessel.
(c) The size of the loft hatch.
(d) Number of occupants using the building.

(54) A room sealed (balanced flued) heater draws the air required to support combustion of the fuel from:

(a) The room only in which the heater is fitted.
(b) Outside the building at a point adjacent to the outlet.
(c) Outside the building at a point some distance from the outlet.
(d) An air line.

(55) A steel panel radiator connected to a secondary return would:

(a) Act as a heat leak.
(b) Prevent the circulation of water.
(c) Be suitable providing it was chromium plated.
(d) Cause the discoloration of the hot water.

(56) Condensation in a conventional flue is the result of:

(a) Excessive amounts of fuel being used.
(b) Excessive flue gas temperatures.
(c) Excessive cooling of the flue gases.
(d) Insufficient air being supplied to support combustion.

(57) A thermocouple is:

(a) A device which generates a small electric current.
(b) A flame failure device.
(c) A fusible link, designed to break and close a gas valve in the event of a fire.
(d) The point where the pilot flame burns.

(58) Which of the following is the most important task to carry out, when leaving a partially completed gas installation:

(a) Ensure the gas supply is turned off at the mains.
(b) Ensure the safety of all the gas pipe runs, irrespective of whether the gas is in the pipeline or not.
(c) Purge the pipeline of air before leaving.
(d) Inform the client when you should return.

(59) The test point located close to a gas appliance is to enable you to:

(a) Check to see the pressure at the burner.
(b) Check the gas rate.
(c) Take a sample of gas.
(d) Check for earth leakages.

(60) The purpose of a flue terminal is to:

(a) Minimise down draughts.
(b) Filter out any remaining undesirable gases from the products of combustion.
(c) Do away with the need for a down draught diverter.
(d) Provide a neater finish to the flue.

(61) Increasing the loading to a gas governor would:

(a) Decrease the outlet pressure.
(b) Increase the outlet pressure.
(c) Decrease the inlet pressure.
(d) Increase the inlet pressure.

(62) The term gas-to-air ratio refers to the:

(a) Calorific value of the gas.
(b) Relative density of the gas compared to air.
(c) Amount of gas in a volume of air to allow ignition.
(d) Volume of air required to give complete combustion of the fuel.

(63) The products of complete combustion of natural gas are:

(a) Carbon dioxide and carbon monoxide.
(b) Oxygen and carbon monoxide.
(c) Hydrogen and methane.
(d) Water vapour and carbon dioxide.

(64) The recommended fuel supplied to an oil-fired pressure jet burner should be:

(a) Class D 35 second fuel.
(b) Class C2 28 second fuel.
(c) Gas oil.
(d) Red diesel fuel oil.

(65) The term viscosity refers to:

(a) The amount of hydrocarbons in the fuel.
(b) The oil's ability to ignite.
(c) The oil's ability to flow easily.
(d) The fact that the amount of smoke produced is high.

(66) An oil level indicator will be found:

(a) Inside the boiler casing.
(b) On or adjacent to the oil storage tank.
(c) Connected to a constant pressure control.
(d) Fitted on the oil supply line close to the boiler.

10 Assessing Your Knowledge

(67) A constant oil level control is used:

(a) On the supply line to a vaporising burner.
(b) On the supply line to an atomising burner.
(c) On the supply line to a pressure jet burner.
(d) On the supply line to an oil storage tank.

(68) The flue/appliance efficiency is found by:

(a) Reducing the smoke reading to its minimum and taking the temperature of the flue gas.
(b) Calculating the oil consumed over a three-month period.
(c) Referral to the manufacturer's instructions.
(d) Comparing the CO_2% against the flue temperature.

(69) A test for earth continuity is carried out in order to:

(a) Ensure the phase conductor is connected to its correct location.
(b) Ensure all exposed metalwork is connected to a suitable earth.
(c) Check all three phases are located within the outlet socket.
(d) Ensure the temporary bonding wire is working effectively.

(70) A typical example of an electrical circuit wired in series would be:

(a) Christmas tree lights.
(b) The ring main.
(c) The lighting circuit.
(d) A battery.

(71) The rating to which a fuse will blow can be found using the following simple formula:

(a) Volts ÷ amps = ohms.
(b) Volts ÷ watts = amps.
(c) Watts ÷ volts = amps.
(d) Watts × volts = amps.

(72) In running the electrical supply to an appliance a spur outlet refers to:

(a) A socket run from the lighting circuit.
(b) A supply run from the ring main to an isolated socket.
(c) A term used to indicate that a boiler has been fitted with a 13 amp 3 pin plug.
(d) A radial circuit run from the consumer unit.

(73) A solenoid is:

(a) A device which uses the current from one circuit to switch on the current to another circuit.
(b) A device which converts a.c. to d.c.
(c) A device used to hold a valve open.
(d) A device which stores a charge of electrical energy.

(74) The trap to a bath which discharges into a single stack system must have a minimum depth of seal of:

(a) 38 mm.
(b) 50 mm.
(c) 65 mm.
(d) 75 mm.

(75) The test to ensure that the trap seals are maintained in a discharge system is known as:

(a) Performance test.
(b) Bourbon test.
(c) Soundness test.
(d) Hydraulic test.

(76) Compression is most likely to occur in a discharge stack:

(a) When a connection is made within 200 mm of a WC branch.
(b) At the foot of the stack.
(c) When two branch connections are opposite.
(d) When no allowance has been made for expansion.

(77) Self-siphonage is most likely to occur in a trap which has water discharged from a:

(a) Bath.
(b) Basin.
(c) Sink.
(d) WC.

(78) Water which is removed from a trap due to the discharge of water from another appliance, further down the discharge pipe, is called:

(a) Self-siphonage.
(b) Induced siphonage.
(c) Back-siphonage.
(d) Momentum.

(79) The contents of a washdown WC pan are removed by:

(a) Induced siphonage.
(b) Siphonic action.
(c) Compression.
(d) Momentum.

(80) An above-ground sanitary discharge system is required to withstand a soundness test of:

(a) 25 mm water gauge.
(b) 38 mm water gauge.
(c) 100 mm water gauge.
(d) 1.5 m head of water.

(81) A sparge pipe will be found fitted to a:

(a) High or low level WC suite.
(b) Slop sink or hopper.
(c) Bowl urinal.
(d) Slab urinal.

(82) The minimum internal diameter of a drain or private sewer used to convey foul water should be:

(a) 50 mm.
(b) 75 mm.
(c) 100 mm.
(d) 150 mm.

(83) The instrument used to measure the air pressure within a drainage pipe during a test is called a:

(a) Manometer.
(b) 'U' tube.
(c) Viscometer.
(d) Absolute pressure gauge.

(84) The base of a soil discharge stack should be fitted with a:

(a) Duck bend.
(b) Back or side inlet gully.
(c) Long radius bend.
(d) Knuckle bend.

(85) When the main drainage is not connected to a dwelling and a septic tank is being used to receive all the foul water, the type of drainage system to be chosen is known as:

(a) Combined.
(b) Separate.
(c) Partially separate.
(d) Surface disposal.

(86) To prevent foul water drains surcharging during heavy rainfall, which of the following systems of drainage should be chosen?

(a) Combined.
(b) Separate.
(c) Relief.
(d) Pumped.

(87) A gusset is inserted into the corner of non-metallic sheet by the formation of a:

(a) Lap joint.
(b) Dog ear.
(c) Delaminated joint.
(d) Split seam joint.

(88) A drip is provided to lead-lined gutters in order to:

(a) Permit thermal movement.
(b) Make fixing easier.
(c) Slow down the flow of water.
(d) Make it possible to achieve the required fall.

(89) When a flat roof is to be covered in sheet copper, where foot traffic can be expected, which of the following expansion joints should not be recommended?

(a) Standing seams.
(b) Batten rolls.
(c) Double lock welts.
(d) Drips.

(90) When flat lap welding to sheet lead the weld should consist of:

(a) One loading of filler metal.
(b) Two loadings of filler metal.
(c) Three loadings of filler metal.
(d) Four loadings of filler metal.

(91) The colour code to identify code 5 sheet lead is:

(a) Green.
(b) Red.
(c) Blue.
(d) Black.

(92) A weathering slate piece is used for:

(a) Weathering a chimney.
(b) Weathering a pipe as it passes through the roof.
(c) Terminating the top edge of a sheet roof covering.
(d) Replacing a broken tile on a roof.

(93) The recommended maximum length of a sheet lead flashing is:

(a) 1 m.
(b) 1.5 m.
(c) 2 m.
(d) 2.5 m.

(94) The SI unit for calorific value is measured in:

(a) Megajoules per kilogram.
(b) Kilograms per cubic metre.
(c) Kilowatts per second.
(d) Kilowatts per hour.

(95) The formula used to find the cross-sectional area of the end of a cylinder is:

(a) π d.
(b) π r².
(c) 2 πr.
(d) π d².

(96) Given that the coefficient of linear expansion of copper is 0.000016, what would the increase in length of a 150 m long copper pipeline be when subjected to an increase in temperature of 10°C?

(a) 0.2 m.
(b) 0.24 m.
(c) 2 mm.
(d) 24 mm.

(97) How many litres would be contained in a tank with a volume of 2.76 m³?

(a) 27.6.
(b) 276.
(c) 2760.
(d) 27 600.

(98) The intensity of pressure acting at a boiler base 3 m below the water level in a feed cistern is:

(a) 29.43 kN/m².
(b) 29.43 kN/m³.
(c) 29.43 N/m².
(d) 29.43 N.

(99) A tank measures 1 m high with a base of 2 m × 0.5 m; the tank itself weighs 10 kg. What is the total weight of the tank and its contents when completely filled with water?

(a) 110 kg.
(b) 1010 kg.
(c) 1100 kg.
(d) 1110 kg.

(100) The quantity of heat required to raise the temperature of 1 litre of water 1°C is:

(a) 1 BTU.
(b) 1 therm.
(c) 4.186 kJ.
(d) 4.2 kW.

Supplementary Assessment

(100 short answer questions)

Special Note to College Lecturers

Answer books (providing answers to these supplementary questions) are available to college and training centres. These may be purchased by sending a cheque for £3.50 (inc. postage and packing) made payable to R. Treloar at the following address: School of Construction, Colchester Institute, Sheepen Road, Colchester, Essex, CO3 3LL.

(1) The safety signs illustrated in Figure 1 indicate things the operative must or must not do. Identify the meaning of each sign.

A	
B	
C	

Ⓐ   Ⓑ   Ⓒ

Figure 1

(2) State how warning signs differ from prohibition and mandatory signs.

(3) Identify the statutory document which must be observed on site at all times to ensure safety.

(4) Identify the contents of each of the fire extinguishers shown in Figure 2 and give examples of their fire preventive uses.

Colour	Contents	Uses (i.e. type of fire)

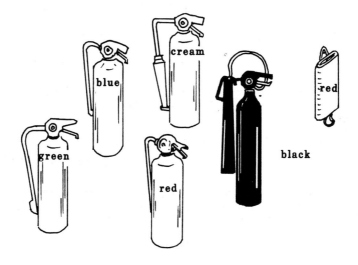

Figure 2

(5) Identify the statutory document which should be observed when carrying out work involving the following:

Water Supply	
Gas Supply	
Electricity	
Building	

(6) State the contents of the series of pipes which have been colour-banded, as shown in Figure 3, found in a boiler house running along a wall.

Pipe No	Contents
1	
2	
3	
4	
5	

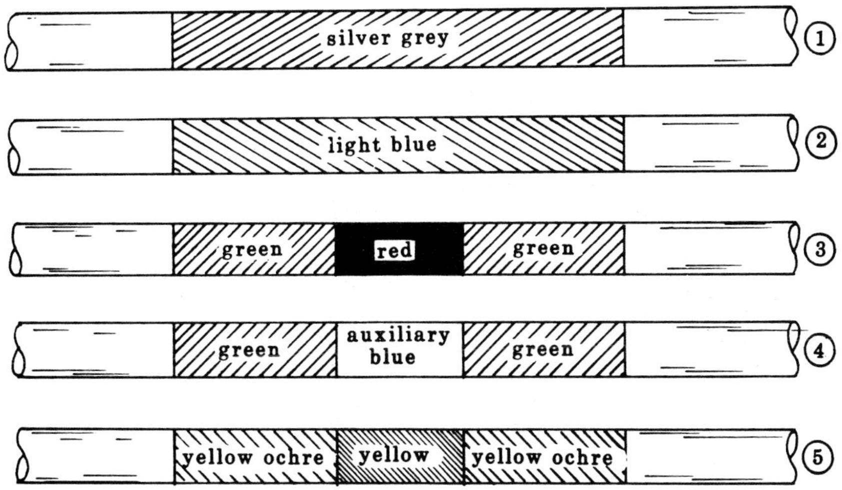

silver grey ①

light blue ②

green | red | green ③

green | auxiliary blue | green ④

yellow ochre | yellow | yellow ochre ⑤

Figure 3

(7) Convert the following SI units into their British Imperial equivalents.

14 kg =	_____ lb
50 litres =	_____ gal

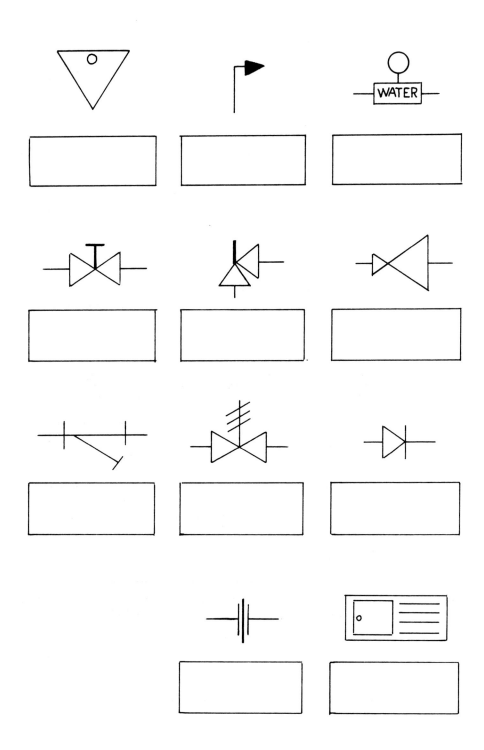

Figure 4

(8) In the space provided below each symbol, indicate the name given to the graphic symbols shown in Figure 4.

(9) Convert the following British Imperial units into their SI equivalents.

5 yds =	_____ m
60 000 BTU/h =	_____ kW

(10) What special precautions need to be observed when using LPG in or around trenches and drains, and why?

(11) With reference to the following materials, complete the table, listing the materials in order of their correct position within the electromotive series, placing the cathodes above the anodes.
Materials: aluminium, copper, lead, tin and zinc.

Physical properties of typical metals

Material in order of the electro-motive series	Chemical symbol	Melting point	Relative density	Specific heat	Coefficient of expansion

(12) Calculate (in mm) the amount of expansion that would occur in a length of copper pipe 30 m long when subjected to a temperature rise of 62°C.

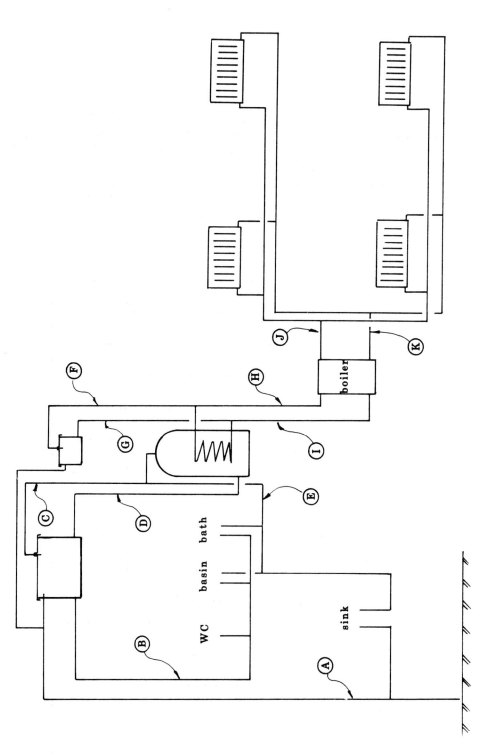

Figure 5

(13) Identify the name given to the design of systems shown in Figure 5.

Cold water system:
Hot water system:
C.H. system:

(14) Complete Figure 5 (using correct graphic symbols) by indicating all the necessary valves, etc., which would be included in the design of the various systems. Also complete the following schedule to identify the name given to each pipe indicated and give a suggested pipe size.

Section	Name	Pipe size
A		
B		
C		
D		
E		
F		
G		
H		
I		
J		
K		

(15) Define what is meant by relative density and specific heat.

Relative density:

Specific heat:

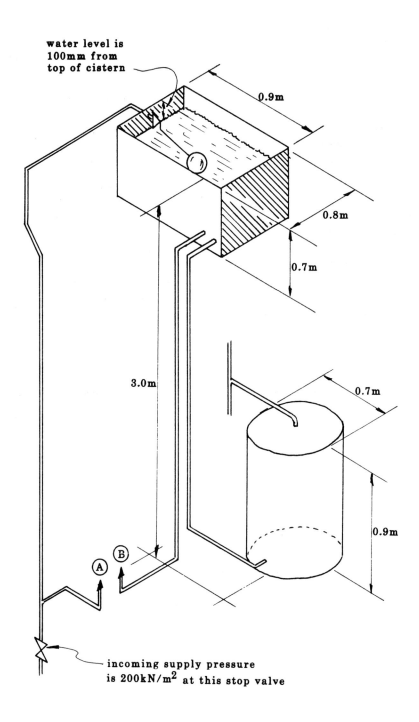

water level is
100mm from
top of cistern

0.9m

0.8m

0.7m

3.0m

0.7m

0.9m

(A) (B)

incoming supply pressure
is 200kN/m^2 at this stop valve

Figure 6

(16) With reference to the dimensions given in Figure 6, calculate the actual capacity of the storage cistern and the capacity of the dhw storage cylinder.

Actual storage capacity of cistern	

Storage capacity of dhw cylinder	

(17) With reference to the dimensions given in Figure 6, calculate the intensity of pressure at tap B and compare this to the mains pressure at tap A. Also find the height to which the mains pressure would rise in a vertical pipe above the stop valve, ignoring the frictional resistance.

Pressure at tap B =

Height to which mains pressure would rise above the stop valve	

(18) Find the total pressure acting upon the base of the dhw cylinder in Figure 6.

(19) In the space provided illustrate a flat dresser.

(20) Give the name of each of the two pipe cutters, illustrated in Figure 7, used to cut low carbon steel tube. Each has an advantage over the other. State what this advantage is.

Cutter A

Cutter B

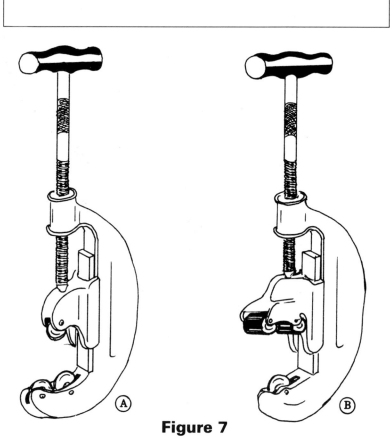

Figure 7

(21) When soft soldering why is the application of a flux required?

(22) Figure 8 illustrates oxyacetylene welding equipment. In the spaces provided round the figure, identify the name of the components indicated.

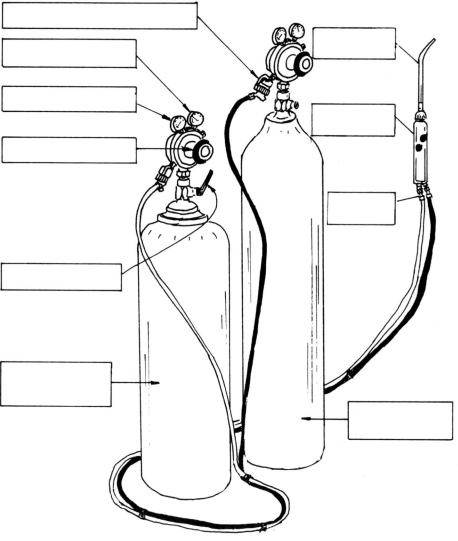

Figure 8

(23) When autogenous welding materials such as lead and steel no flux is required. State why this is so.

(24) Figure 9 shows two completed bronze welded joints. In the spaces provided near the figure give the name used to describe each joint.

(25) Although the joints indicated in question 24 are referred to as welded joints they

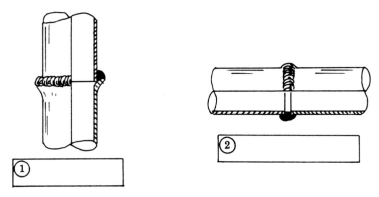

Figure 9

are not truly welded. State why this is so and identify what type of joints they are.

(26) Shown in Figure 10 is a section through a screwdown valve. State the name given to the tap illustrated and in the spaces provided name the arrowed components.

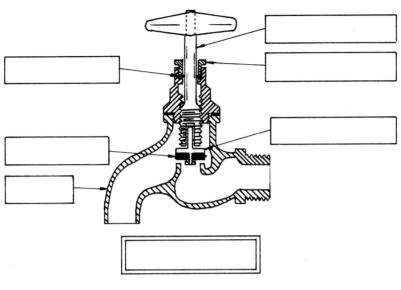

Figure 10

(27) State why a gate valve is fitted in the cold feed pipe to a vented system of dhw in preference to a stopcock.

(28) Under current local water byelaws only float-operated valves conforming to BS 1212, Parts 2 or 3, may be fitted to cisterns in domestic premises; give the reasons for this.

(29) Complete Figure 11 using correct graphic symbols to show an indirect system of cold water supply.

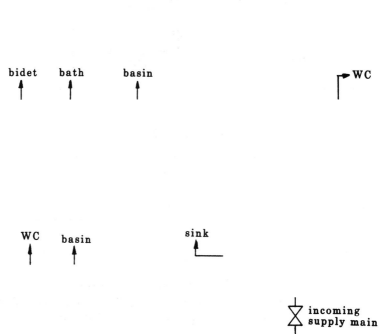

Figure 11

(30) Illustrate in the space provided how two small cisterns may be coupled together in a domestic situation to give a larger capacity, ensuring undue stagnation of the water will not occur.

(31) With reference to Figure 12 name the components, numbered 1–6, fitted to the cold supply of the unvented system of dhw.

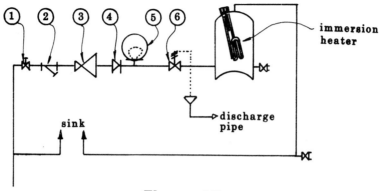

Figure 12

1.	4.
2.	5,
3.	6.

(32) Name the valve missing from Figure 12 which ensures complete safety regardless of excessively rising water temperatures caused by thermostat failure. (Without this device the system does not comply with the Building Regulations.)

(33) Each of the dhw systems illustrated in Figure 13 (over) shows major errors; state what these faults or contraventions are.

System A

System B

System C

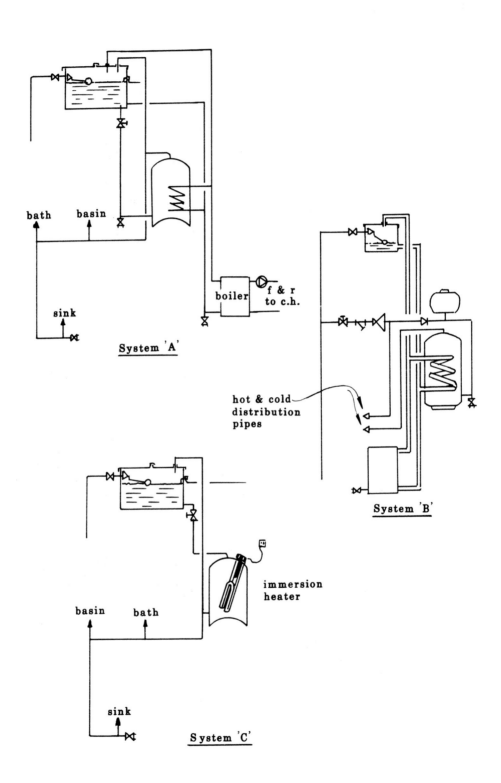

bath   basin

sink

System 'A'

boiler   f & r
to c.h.

hot & cold
distribution
pipes

System 'B'

immersion
heater

basin   bath

sink

System 'C'

Figure 13

(34) Name the instantaneous water heater shown in Figure 14 and state the purpose of the valve indicated at 'X'.

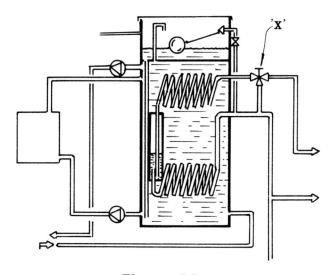

Figure 14

System referred to as:
Purpose of valve 'X':

(35) With reference to Figure 15, indicate in the space provided the maximum permissible depth which can be cut into the floor joist. Also indicate the maximum permissible distance at which the notch can be cut from the supporting wall.

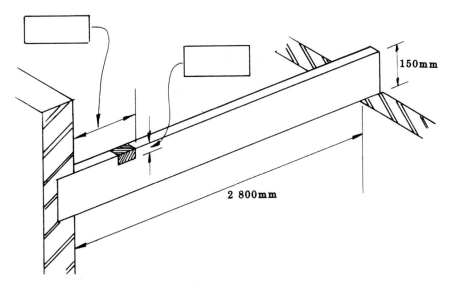

Figure 15

(36) Complete the illustration in Figure 16 to show:

    (a) A one-pipe wet c.h. system.

    (b) A two-pipe wet c.h. system.

    (c) A two-pipe reversed return wet c.h. system.

(37) State why the water level is adjusted low down in the f & e cistern feeding a wet c.h. system.

(38) With reference to the bungalow illustrated on page 133 under the entry of Radiator and Boiler Sizing, complete the table below to find the heat emitter requirements for the bathroom.

Heat requirements. Location: bathroom

Fabric loss Element	Area L × B = (m²)	Temp. diff. (°C)	U value (W/m²°C)	Heat loss (W)
Window	×	×		
External walls	×	×		
Internal walls	×	×		
Floor	×	×		
Roof	×	×		

Ventilation loss
     volume × air change × temp. diff. × factor
 	
                                                 =
     Plus 15% for intermittent heating =				

(39) What gas is produced as a result of incomplete combustion of natural gas?

(40) What do the initials LPG stand for?

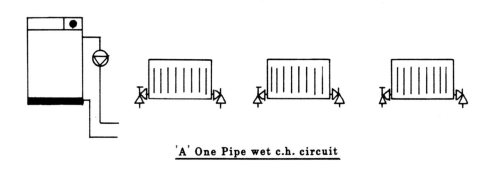

'A' One Pipe wet c.h. circuit

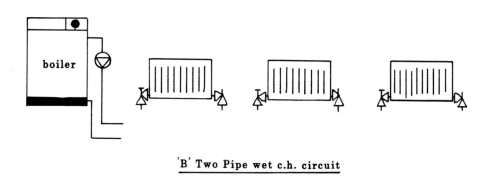

'B' Two Pipe wet c.h. circuit

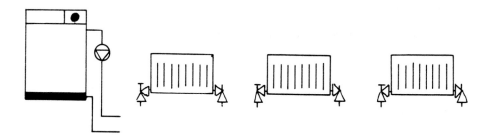

'C' Two Pipe Reversed Return wet c.h. circuit

Figure 16

(41) When carrying out a soundness test to an existing natural gas installation what is the pressure (in mbar) to which the system should be tested?

(42) What is meant by the term let-by when testing gas installations? Describe how this test is achieved.

(43) What is the name given to the device fitted in the pipeline to a gas appliance which maintains the gas at a constant pressure as recommended by the manufacturer?

(44) Complete the illustration at Figure 17 to show the flue pipe, terminal and minimum distances to be observed when terminating the natural draught open flue from a natural gas appliance.

(45) What is meant by the term 'room sealed appliance' and why should these appliances generally be fitted in preference to open-flued appliances?

(46) Calculate the effective free air ventilation requirements for an open-flued appliance of 18 kW natural gas input.

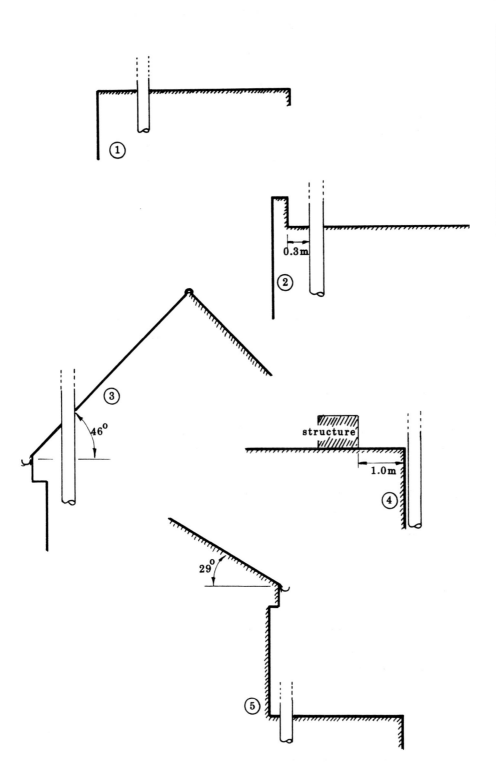

Figure 17

(47) With reference to the air grille illustrated in Figure 18, calculate its effective free air size.

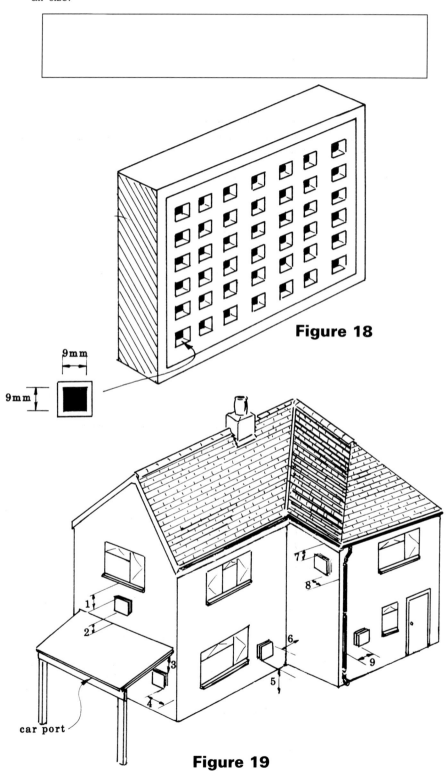

Figure 18

9mm

9mm

car port

Figure 19

(48) Identify the minimum terminal location dimensions indicated for the natural draught balanced flue appliances shown in Figure 19.

1.	4.	7.
2.	5,	8.
3.	6.	9.

(49) State the minimum ventilation requirements (in mm²) for a decorative fuel effect gas fire of 6 kW input rating into a room.

(50) Two grades of fuel are used for domestic oil fired burners; name these and explain how the fuels differ.

(51) With reference to Figure 20 complete the illustration of the gravity-fed one-pipe oil supply system to show the necessary components and controls to be located on the storage tank and pipeline to enable its safe, efficient and effective use.

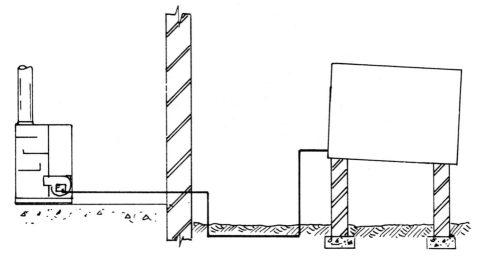

Figure 20

(52) Complete the illustration of the 'fusible link' type fire valve, as shown in Figure 21, and explain its operation.

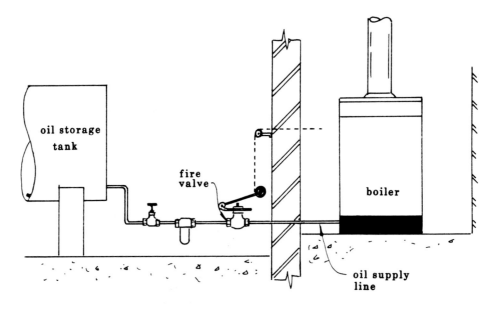

Figure 21

Operation of fire valve:

(53) State the purpose of a constant oil level control and indicate where one would be found.

(54) Use the spaces provided to name the arrowed components on the pressure jet burner shown in Figure 22.

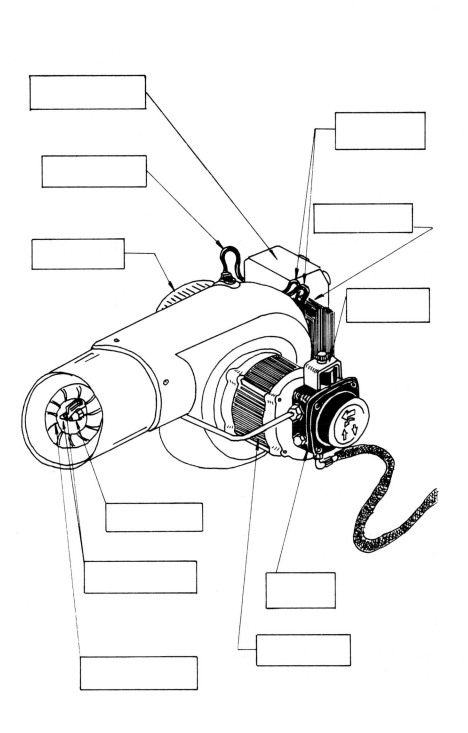

Figure 22

(55) Identify the device located in a pressure jet burner to detect light rays from within the combustion chamber to confirm ignition of the fuel and as a result making or breaking an electrical circuit.

```
┌─────────────────────────────────────────────────────────────┐
│                                                               │
│                                                               │
└─────────────────────────────────────────────────────────────┘
```

(56) Calculate the free air ventilation requirements for an open-flued pressure jet burner with an input rating of 18 kW.

```
┌─────────────────────────────────────────────────────────────┐
│                                                               │
│                                                               │
│                                                               │
│                                                               │
└─────────────────────────────────────────────────────────────┘
```

(57) When should a draught stabiliser be fitted to the flue way of an open-flued oil burning appliance?

```
┌─────────────────────────────────────────────────────────────┐
│                                                               │
│                                                               │
│                                                               │
└─────────────────────────────────────────────────────────────┘
```

(58) Name the four separate tests which are carried out when completing a combustion efficiency test to an oil burning appliance. Also indicate with an 'X' the test which is not applicable to balanced flued appliances.

1.	
2.	
3.	
4.	

(59) Show the correct location of the fuse to be fitted into the electrical circuit shown in Figure 23. Also calculate the size of fuse to be fitted into the 13 amp 3-pin plug of an electric drill with a power rating of 550 W, designed to run on 240 V.

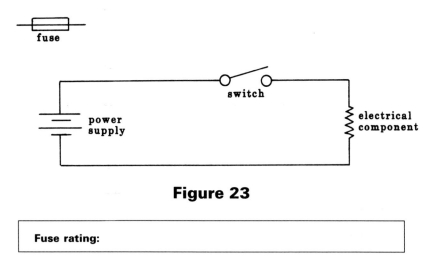

Figure 23

Fuse rating:

(60) Complete the two wiring diagrams shown in Figure 24 to produce one circuit in series and one in parallel.

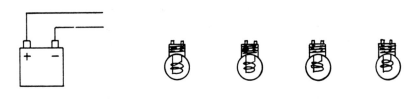

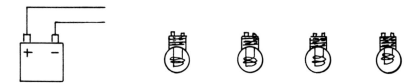

Figure 24

(61) What is the common name given to the circuit protective conductor found in domestic house wiring?

(62) When would a temporary bonding wire be used?

(63) When replacing a ceramic Butler sink with a new stainless steel sink top what responsibility has the plumber to the client with regard to electrical safety?

(64) Figure 25 shows a completed ring circuit; show how the wiring can be altered to accommodate the connection of the spur outlet.

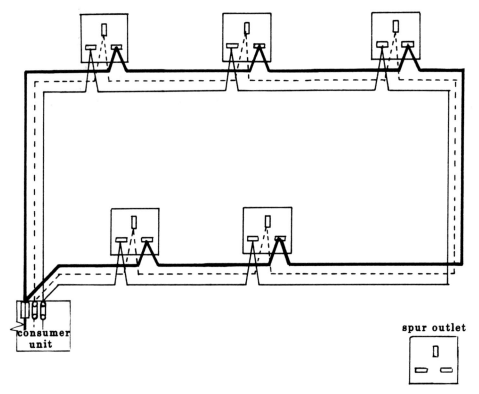

Figure 25

(65) Using coloured pencils, or the key which identifies the colour of cables, complete Figure 26 to indicate the location of the brown, blue, and green and yellow conductors into the 13 amp 3-pin plug.

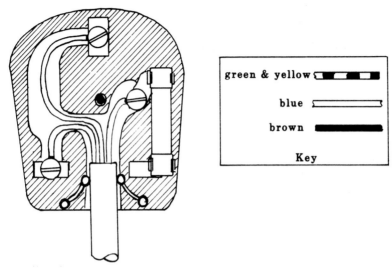

Figure 26

(66) How should cables be protected from damage when passing through metalwork?

(67) Using a simple illustration explain the operation of the rod thermostat found inside an immersion heater to a dhw storage cylinder.

(68) A transformer has half the number of coils on the secondary windings as those on the primary coil. Is it a step up or step down transformer?

(69) Figure 27 shows the top of an immersion heater with the cover removed; using coloured pencils (or the key in Figure 26) complete the illustration to show the electrical connections to the heater element. Also, in the space provided, state the size and type of flex to be used.

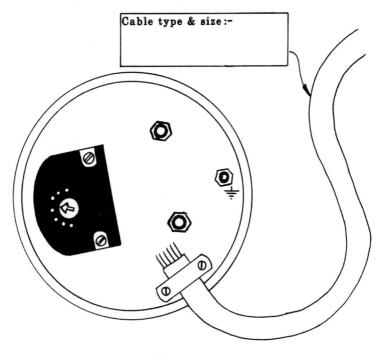

Cable type & size :-

Figure 27

(70) State the effective size for a window opening which is to be fitted for the purpose of ventilation to a room used for sanitary accommodation.

(71) Figure 28 shows a trap. In the space provided give the name of the trap design and indicate the depth of water seal.

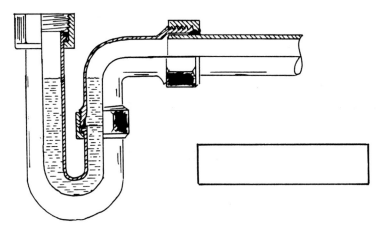

Figure 28

(72) Complete the table below to indicate the minimum size of waste fitting, trap and discharge pipe size.

Type of appliance	Size of waste fitting	Discharge pipe and trap size
Wash basin		
Sink		
Bath		
Bidet		
Shower tray		
Bowl urinal		
Stall urinal		
Drinking fountain		

(73) With reference to Figure 29, in the spaces provided, give the minimum distances which need to be maintained for the proposed discharge pipes in order to avoid cross-flow of effluent from one discharge pipe into another.

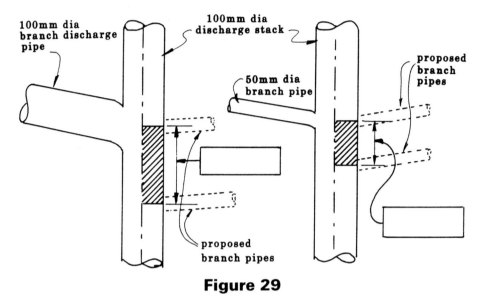

100mm dia branch discharge pipe

100mm dia discharge stack

50mm dia branch pipe

proposed branch pipes

proposed branch pipes

Figure 29

(74) Name the system of sanitary pipework shown in Figure 30, where no additional ventilation pipework is required. Also indicate in the space provided the maximum dimensions to be observed.

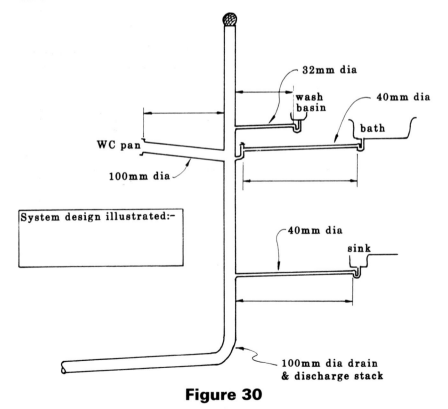

32mm dia

wash basin

40mm dia

bath

WC pan

100mm dia

System design illustrated:-

40mm dia

sink

100mm dia drain & discharge stack

Figure 30

(75) Complete the following table to give the maximum number of appliances to be installed within a single unvented branch discharge pipe.

Unvented branch discharge pipes serving more than one appliance

Appliance	Maximum No of appliances to be fitted	Minimum pipe diameter
Wash basins		
Bowl urinals		
WC pans		

(76) What is the purpose of a resealing trap?

(77) What is meant by the term 'induced siphonage'?

(78) Complete the illustration in Figure 31 (over) to show how an air test is maintained in above-ground soundness testing. Also state the minimum air test pressure.

Minimum air pressure:

(79) State what is meant by the term 'performance test'.

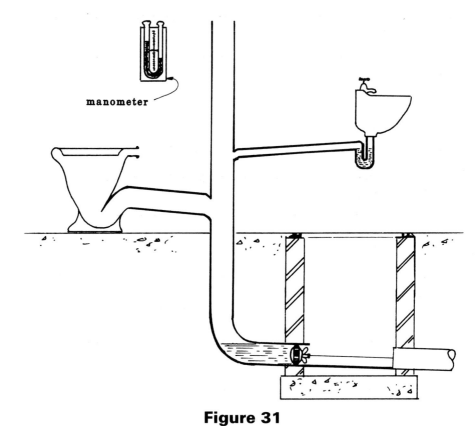

manometer

Figure 31

(80) Complete Figure 32 to show a partially separate system of drainage.

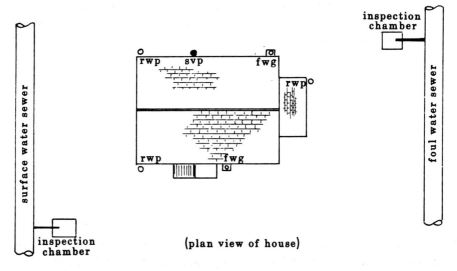

surface water sewer

inspection chamber

inspection chamber

foul water sewer

rwp   svp   fwg

rwp

rwp   fwg

(plan view of house)

Figure 32

(81) Identify the purpose of an anti-flood gully and state the type of drainage system in which it may be found.

```



```

(82) Give four different possible locations where a point of access will be required to a system of drainage.

1.
2.
3.
4.

(83) What is meant by the term 'benching'?

```



```

(84) Show by illustration how a drainage pipe run can be made to pass through a wall, or foundation, below ground level to ensure the pipe will not fracture due to movement.

```



```

(85) What is the name given to a piece of wood cut to an angle and used in conjunction with a spirit level for setting out the gradient to a short drainage run?

```


```

(86) Complete Figure 33 to show the material, and indicate the depths of bedding material to be used when laying the 100 mm uPVC drainage pipe into the ground.

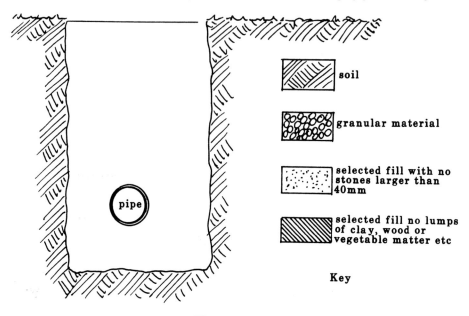

Figure 33

(87) Complete Figure 34 to show how the uPVC drain at high level can be made to connect suitably to the drain at the lower level.

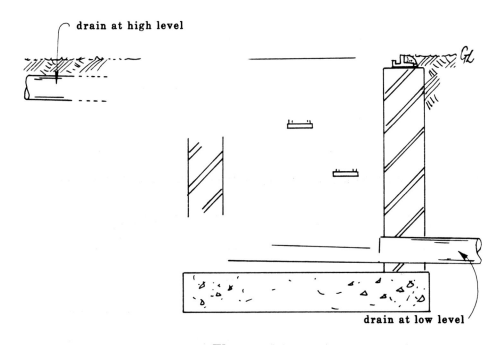

Figure 34

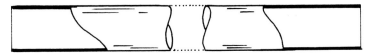

Figure 35

(88) Complete Figure 35 to show how a 1.5 m minimum water head pressure test can be achieved to the drainage pipe.

(89) With reference to the previous question state the maximum water loss permitted from the pipe over a 30 min test period.

90) State the purpose of a soakaway.

(91) Complete the following table, giving the British Standard specifications for sheet lead.

BS 1178 Code No	Colour code	Thickness (mm)
3		
4		
5		
6		

(92) In the space provided name the chimney flashing details shown in Figure 36; also give the calculation used to determine the length of a soaker.

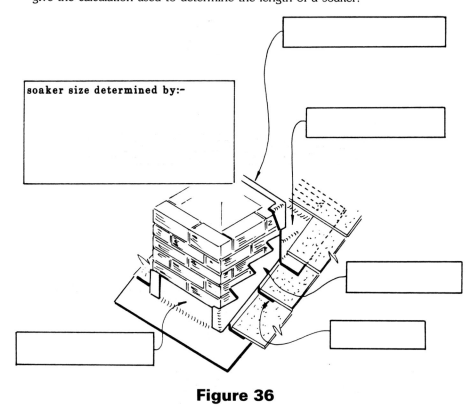

soaker size determined by:-

Figure 36

(93) State the maximum length to be observed for any flashing detail.

(94) Figures 37 and 38 show a wood-cored roll and drip, as used for sheet lead. Complete the illustrations to show how the lead is weathered at these details, giving dimensions and explanatory notes as necessary.

(95) State the name of the fixing used to secure a cover flashing into the brick course.

(96) Given that code 5 sheet lead is going to be used to weather a small gutter lining, with a distance of 600 mm between the wood-cored rolls, state the maximum length to be observed between drips.

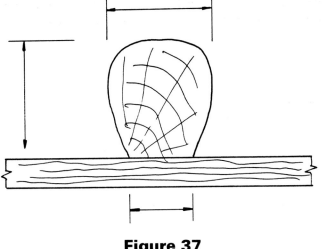

Figure 37

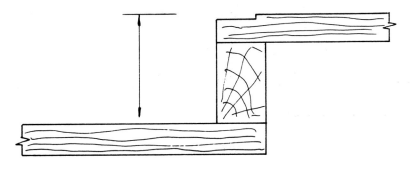

Figure 38

(97) Name the weathering detail shown in Figure 39.

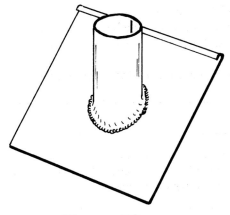

Figure 39

(98) State the minimum vertical cover that needs to be maintained with any lap joint, such as at cover flashings.

(99) State the recommended finish height for a standing seam as used for copper and aluminium roof coverings.

(100) Figure 40 shows a batten roll as used for copper and aluminium roof coverings. Complete the illustration to show how the sheet material is weathered at this detail, giving dimensions and explanatory notes as necessary.

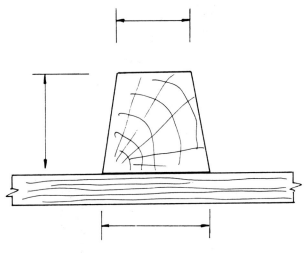

Figure 40

Problem Solving

Problem No 1

The f & e cistern of the system illustrated below is continually discharging water from the overflow. The maintenance plumber who was called initially replaced the float-operated valve in the cistern; however, he was called back to the job to re-investigate. On this occasion he turned off the service valve feeding the system and switched off the boiler. He now emptied all the excess water from the cistern but within 10 minutes the cistern started to overflow again. What is the cause?

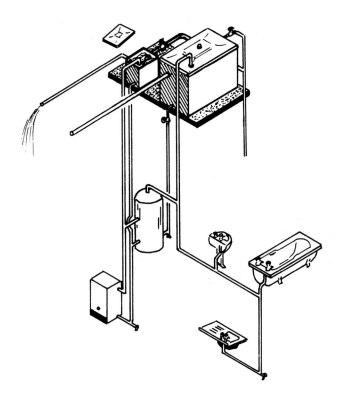

Answer:

Problem No 2

You have been called to a job to investigate the problem of a poor and intermittent flow of water from tap 'X' when the other taps are being used. State the cause and remedy.

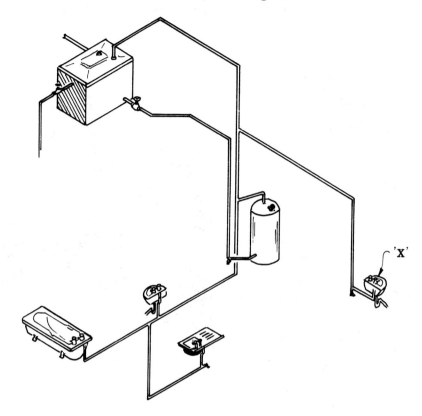

Answer:

Problem No 3

During the summer months the fully-pumped dhw and c.h. system illustrated below is faulty in that the heat emitters at A and B heat up when the system is operating on dhw only. There is no fault with the wiring and the electrical controls are working correctly. Identify the fault.

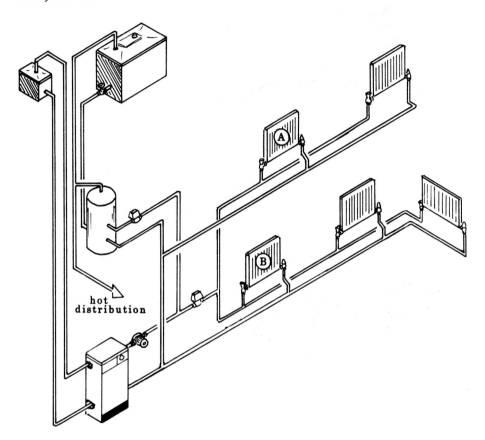

hot
distribution

Answer:

Problem No 4

The fully pumped heating system shown below suffers the problem of air continually being drawn into the system; this invariably causes the heat emitters to become cold and as a result require frequent venting. State the cause of this problem and suggest a remedy. Also identify the long-term effects this problem will have on the system.

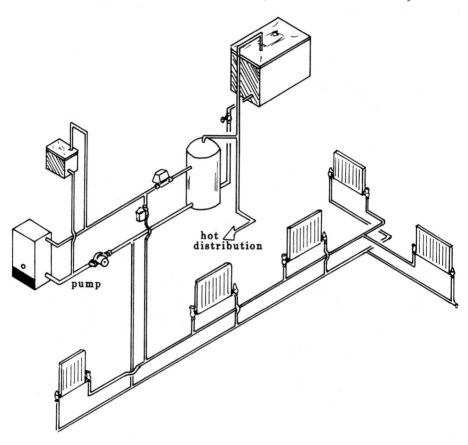

hot distribution

pump

Answer:

Problem No 5

You have been called to look at the dhw system illustrated below to investigate the problem of brown water which keeps appearing at the hot draw-off taps. The householder informs you that the c.h. system has recently been extended by the addition of three large radiators and that the problem did not occur before this time; the system itself was installed some 15 years ago. What is the likely cause and remedy?

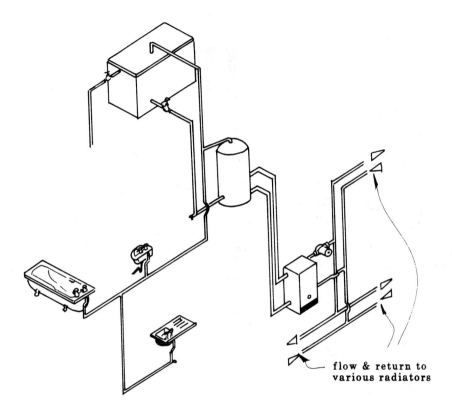

flow & return to
various radiators

Answer:

Problem No 6

You have been called to the plumbing system illustrated below. The householder says that the dhw seems to take 6–8 hours to warm up, yet the boiler is working fine, although it is noisy at times. State the likely cause, remedy and work required to prevent this problem's reoccurrence.

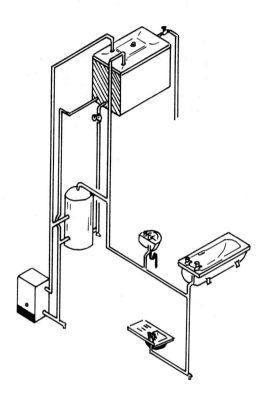

Answer:

Problem No 7

You have been called to the gas installation shown below because the 5 kW (17 000 btu/h) natural gas fire does not seem to operate effectively all the time and gives out very little heat; basically the flames seem to burn very low. Investigation of the flue has revealed no fault and there is a good pull to the chimney. The fire itself has no apparent defects. State the probable cause and suggest its remedy.

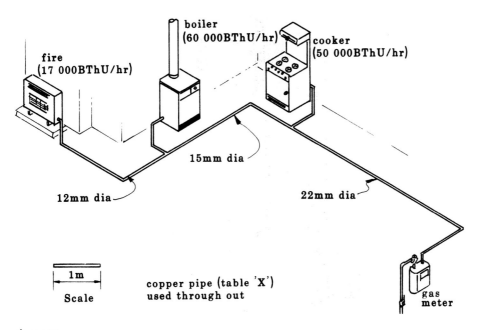

Answer:

Problem No 8

Illustrated below is an unvented dhw unit which is experiencing intermittent discharge of water from the pressure relief valve. State the possible cause of the problem and how it can be cured. Also, the client is worried that no drain-off cock has been fitted to the base of the dhw cylinder. Note that it is fitted to the top of the dhw storage vessel; explain how the water is removed from the vessel if necessary.

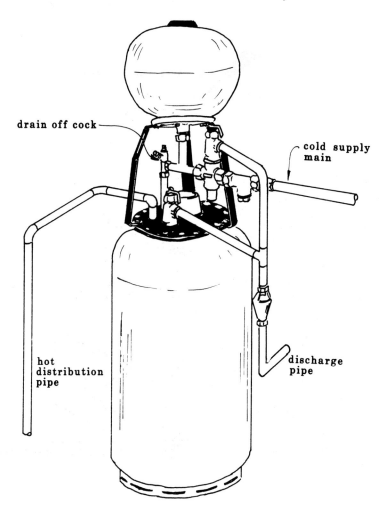

drain off cock

cold supply main

hot distribution pipe

discharge pipe

Answer:

Problem No 9

You have been called to a dwelling to investigate the fatality of an occupant (found in the extension) owing to carbon monoxide poisoning. A 5 000 mm² permanent air grille is installed in the room where a 17 kW open-flued natural gas burning appliance has been installed. A test on the flue system indicates no signs of spillage at the draught diverter. The appliance had been installed some 8 months previously and had been working most satisfactorily until the accident. What fault in the installation caused the death of the occupant?

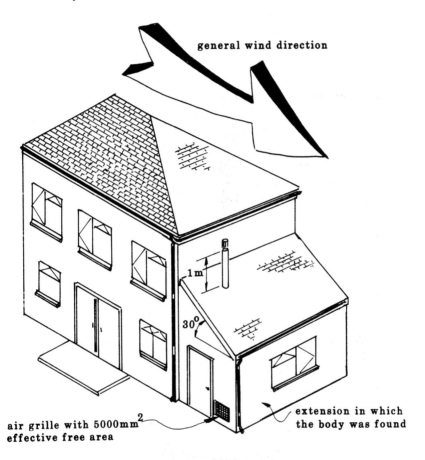

general wind direction

1 m

30°

air grille with 5000mm² effective free area

extension in which the body was found

Answer:

Problem No 10

Having just completed the installation of the indirect cold water system, illustrated below, and turned on the water supply, you find that water flows freely from the bath yet only trickles out of the taps to the wash basin and WC cistern. What is the likely cause of this problem?

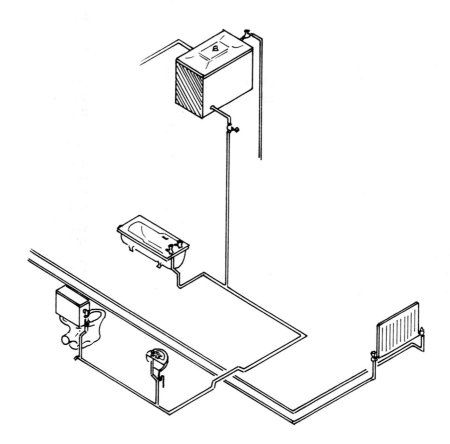

Answer:

Answers to Multiple Choice Questions

(1) a	(21) a	(41) b	(61) b	(81) d
(2) c	(22) a	(42) b	(62) c	(82) c
(3) c	(23) a	(43) c	(63) d	(83) a
(4) c	(24) d	(44) b	(64) b	(84) c
(5) a	(25) c	(45) a	(65) c	(85) b
(6) a	(26) d	(46) b	(66) b	(86) b
(7) b	(27) a	(47) a	(67) a	(87) c
(8) c	(28) c	(48) a	(68) d	(88) a
(9) c	(29) a	(49) b	(69) b	(89) a
(10) a	(30) c	(50) c	(70) a	(90) b
(11) b	(31) a	(51) a	(71) c	(91) b
(12) a	(32) a	(52) d	(72) b	(92) b
(13) b	(33) b	(53) a	(73) c	(93) b
(14) c	(34) d	(54) b	(74) d	(94) a
(15) a	(35) c	(55) d	(75) a	(95) b
(16) d	(36) b	(56) c	(76) b	(96) d
(17) c	(37) c	(57) a	(77) b	(97) c
(18) b	(38) c	(58) b	(78) b	(98) a
(19) c	(39) a	(59) a	(79) d	(99) b
(20) c	(40) c	(60) a	(80) b	(100) c

Index

Index